PRICE LEVEL MEASUREMENT

CONTRIBUTIONS
TO
ECONOMIC ANALYSIS

196

Honorary Editor:
J. TINBERGEN

Editors:
D. W. JORGENSON
J. WAELBROECK

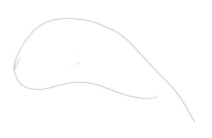

NORTH-HOLLAND
AMSTERDAM • NEW YORK • OXFORD • TOKYO

PRICE LEVEL MEASUREMENT

Edited by

W.E. DIEWERT

Department of Economics
The University of British Columbia
Vancouver, Canada

1990

NORTH-HOLLAND
AMSTERDAM • NEW YORK • OXFORD • TOKYO

ELSEVIER SCIENCE PUBLISHERS B.V.
Sara Burgerhartstraat 25
P.O. Box 211, 1000 AE Amsterdam, The Netherlands

Sole distributors for the United States:

ELSEVIER SCIENCE PUBLISHING COMPANY, INC.
655 Avenue of the Americas
New York, N.Y. 10010, U.S.A.

ISBN: 0 444 88108 5

PRINTED IN THE NETHERLANDS

INTRODUCTION TO THE SERIES

This series consists of a number of hitherto unpublished studies, which are introduced by the editors in the belief that they represent fresh contributions to economic science.

The term "economic analysis" as used in the title of the series has been adopted because it covers both the activities of the theoretical economist and the research worker.

Although the analytical methods used by the various contributors are not the same, they are nevertheless conditioned by the common origin of their studies, namely theoretical problems encountered in practical research. Since for this reason, business cycle research and national accounting, research work on behalf of economic policy, and problems of planning are the main sources of the subjects dealt with, they necessarily determine the manner of approach adopted by the authors. Their methods tend to be "practical" in the sense of not being too far remote from application to actual economic conditions. In additon they are quantitative.

It is the hope of the editors that the publication of these studies will help to stimulate the exchange of scientific information and to reinforce international cooperation in the field of economics.

The Editors

TABLE OF CONTENTS

PRICE LEVEL MEASUREMENT – W.E. Diewert (Editor)
Canadian Government Publishing Centre /
Elsevier Science Publishers B.V. (North-Holland)
© Minister of Supply and Services Canada, 1990

AN OVERVIEW OF THE PAPERS ON PRICE LEVEL MEASUREMENT

W.E. Diewert

The first theoretical paper is by Pollak, "The Theory of the Cost-of-Living Index". This paper was originally issued by the U.S. Bureau of Labor Statistics as BLS Working Paper 11 in June 1971. The paper has been extremely influential and is regarded as a "classic" in the field by researchers in index number theory. Pollak not only develops the theory of the cost-of-living index which the CPI approximates, he also studies the related problem of measuring welfare. Some of the contributions that Pollak makes in this paper are: (i) he provides a systematic theory of empirically implementable bounds for the single consumer cost-of-living index, (ii) he systematically develops the theory of the Malmquist [1953] quantity or welfare index, and (iii) he brings out the connection between functional forms for a consumer's welfare or utility function and the functional form for the corresponding cost-of-living index.

The paper by Diewert, "The Theory of the Cost-of-Living Index and the Measurement of Welfare Change", continues to develop the theoretical foundations of CPI. Drawing on Pollak [1981], Diewert develops the many- consumer theory of the cost-of-living index. Diewert adapts a technique due originally to Konüs [1924] and shows that many theoretically unobservable indexes may be bounded by empirically computable Paasche and Laspeyres indexes. He also discusses several other issues from a theoretical perspective, including: (i) the use of the chain principle versus the fixed base principle in constructing index numbers, (ii) the problems involved in constructing group welfare indexes, (iii) the construction of subindexes and how they may be combined to form an overall index and (iv) the treatment of housing and other consumer durables in the CPI.

The paper by Jorgenson and Slesnick, **"Individual and Social cost of Living Indexes"**, is an intellectual tour de force as Russell notes in his perceptive comments on the paper. It is not possible to summarize accurately the entire paper in this brief introduction; however, some of the major ideas in the paper can be mentioned here. The core of the Jorgenson-Slesnick paper is an econometric model of aggregate consumer behavior which is based on Lau's [1977] theory of exact aggregation which in turn (greatly) generalizes the aggregate model of consumer behavior implemented by Berndt, Darrough and Diewert [1977]. The basic idea is that each household is assumed to have the same preferences (apart from a stochastic term) once we standardize for various demographic characteristics or attributes such as family size, age of head, region of residence, race, rural versus urban, etc. By shrewdly choosing the functional form for the household preferences, Jorgenson and Slesnick end up with a system of aggregate consumer demand functions that depend on consumer prices, the distribution of income, and the distribution of household characteristics. Given aggregate time series and some cross section data the parameters which characterize the household preferences may be estimated econometrically and this is done in Jorgenson, Lau and Stoker [1982]. Once household preferences are known, individual cost-of-living or welfare indexes may be calculated as well as aggregate indexes. Jorgenson and Slesnick also compare their "econometric" approach to the construction of aggregate cost-of-living and social welfare indexes to the "exact" index number approach outlined in the earlier paper by Diewert. There are advantages and disadvantages to each approach. One disadvantage of the Jorgenson-Slesnick approach is that new time series information would lead to new econometric estimates of the household preference parameters and hence new historical index numbers would have to be calculated. However, there is much to be praised in the Jorgenson-Slesnick approach as Russell notes in his comment.

Riddell in his paper, "Leisure Time and the Measurement of Economic Welfare" extends the traditional theory of the cost-of-living index to incorporate the consumer-worker's labour supply (or leisure choice) decision. He also empirically implements his approach using Canadian per capita data. He compares his approach with the traditional approach (which neglects the labour supply decision) and finds that real income growth is dramatically slower in Canada when the labour supply decision is taken into account. It should also be mentioned that Riddell's approach takes into account income taxes as well as commodity taxes. Thus if the government decreased commodity taxes and increased income taxes in a neutral manner, Riddell's welfare indicator would show no change whereas a "traditional" welfare indicator (such as per capita national income deflated by the CPI) would show an increase. Riddell also extends and empirically implements the Pencavel [1977] – Cleeton [1982] theory of real wage indexes.

In the paper, "Preference Diversity and Aggregate Economic Cost-of-Living Indexes", Blackorby and Donaldson look for conditions on the preferences of individual consumers that are sufficient to ensure that an aggregate cost-of-living index is independent of the distribution of utilities or real incomes of the households in the economy. Their results are interesting but essentially negative: it appears that individual household preferences must be homothetic in order for the aggregate cost-of-living index to be independent of the distribution of utilities. Of course, homothetic preferences must be rejected on empirical grounds since homothetic preferences imply unitary income elasticities for all goods (which contradicts Engel's Law).

Eichhorn and Voeller in their paper, "The Axiomatic Foundations of Price Indexes and Purchasing Power Parities", outline the "axiomatic" or "test" approach to index number theory, as opposed to the "economic" approach which was followed by the earlier papers from Pollak to Blackorby and Donaldson. In his comments on Eichhorn and Voeller, Blackorby explains the difference between the "test" approach and the "economic" approach to a price index: in the former approach, prices and quantities in both periods are taken to be independent variables whereas in the latter approach, quantities are regarded as dependent variables. It should be mentioned that Diewert in his paper "The Theory of Cost-of-Living Index and the Measurement of Welfare Change" provides an axiomatic characterization of the "economic" approach to the cost-of-living index. In defence of Eichhorn and Voeller's approach, it should be noted that the "economic" approach to the cost-of-living index followed by most authors in this volume may not be the correct one. The "economic" approach to index number theory rests on the assumption of utility maximizing behaviour which may or may not be true. If it is not true, then there is a need for an alternative theoretical foundation for the CPI, and the "axiomatic" approach developed by Eichhorn and Voeller can fill this need.

References to papers not in this volume

Berndt, E.R., W.E. Diewert and M.N. Darrough [1977], "Flexible Functional Forms and Expenditure Distributions: An Application to Canadian Consumer Demand Functions", *International Economic Review* 18, pp.651-675.

Cleeton, D.L. [1982], "The Theory of Real Wage Indices", *American Economic Review* 72, pp.214-225.

Lau, L.J. [1977], "Existence Conditions for Aggregate Demand Functions", Technical Report No. 248, Institute for Mathematical Studies in the Social Sciences, Stanford University, Stanford, California.

Jorgenson, D.W., L.J. Lau and T.M. Stoker [1982], "The Transcendental Logarithmic Model of Aggregate Consumer Behavior", pp.97-238 in *Advances in Econometrics*, Vol. 1, R.L. Basmann and G.F. Rhodes Jr. (eds.), Greenwich: JAI Press.

Konüs, A.A. [1924], "The Problem of the True Index of the Cost-of-Living", Translated in *Econometrica* 7 [1939], pp.10-29.

Malmquist, S. [1953], "Index Numbers and Indifference Surfaces", *Trabajos de Estadística* 4, pp.209-242.

Pencavel, J.H. [1977], "Constant Utility Index Numbers of Real Wages", *American Economic Review* 67, pp.91-100.

Pollak, R.A. [1975], "Subindexes of the Cost-of-Living Index", *International Economic Review* 16, pp.135-150.

_____ [1981], "The Social Cost-of-Living Index", *Journal of Public Economics* 15, pp.311-336.

PRICE LEVEL MEASUREMENT – W.E. Diewert (Editor)
Canadian Government Publishing Centre /
Elsevier Science Publishers B.V. (North-Holland)
© Minister of Supply and Services Canada, 1990

5

THE THEORY OF THE COST-OF-LIVING INDEX

Robert A. Pollak*
Department of Economics
University of Pennsylvania

SUMMARY

In this paper I summarize the theory of the "cost-of-living index" and the closely related theory of the "preference field quality index". The first section is devoted to notation and preliminary background, and the second summarizes the theory of the cost-of-living index. The next two sections discuss upper and lower bounds for the index, and the preference orderings for which these bounds are attained. In Section 5 I examine the effect on the index of the choice of a base indifference curve. Section 6 examines the form of the cost-of-living index corresponding to various specific preference orderings. In Sections 7 and 8 I examine the preference field quantity index, a quantity index which is conceptually similar to the cost-of-living index. I show that it is equal to the expenditure index deflated by the cost-of-living index if and only if the preference ordering is homothetic to the origin. In Section 9 I discuss the use of price indexes in empirical demand analysis and argue that the cost-of-living index is inappropriate for this purpose.

In this paper I summarize the theory of the "cost-of-living index" and the closely related theory of the "preference field quality index". The first section is devoted to notation and preliminary background, and the second summarizes the theory of the cost-of-living index. The next two sections discuss upper and lower bounds for the index, and the preference orderings for which these bounds are attained. In Section 5 I examine the effect on the index of the choice of a base indifference curve. Section 6 examines the form of the cost-of-living index corresponding to various specific preference orderings. In Sections 7 and 8 I examine the preference field quantity index, a quantity index which is conceptually similar to the cost-of-living index. I show that it is equal to the expenditure index deflated by the cost-of-living index if and only if the preference ordering is homothetic to the origin. In Section 9 I discuss the use of price indexes in empirical demand analysis and argue that the cost-of-living index is inappropriate for this purpose.

There are a number of areas of practical and theoretical importance which I do not discuss in this paper. No mention is made of quality change, goods provided by governments, or the treatment of the environment. Goods are assumed to be perfectly divisible and transaction costs to be zero. Intertemporal aspects of the consumer's allocation problem are ignored, so the treatment of saving, interest rates, and consumer durables is not discussed. I have ignored these areas not only because they are difficult but also because a systematic theoretical attack on any of them must be based on a thorough understanding of the basic theory. A major purpose of this paper is to present a rigorous survey of that theory. The material in the first two sections should be familiar, but much of the material in later sections is new.

The discussion of bounds on the cost-of-living index in Sections 3 and 4 emphasizes the importance of both upper and lower bounds. It is well known that the Laspeyres index provides an upper bound to the cost-of-living index based on the indifference curve attained in the reference situation, but little attention has been given to the corresponding lower bounds. In Section 3 I establish upper and lower bounds, and in Section 4 I characterize the preference orderings for which these bounds are attained.

It is well known that the cost-of-living index is independent of the base indifference curve if and only if the indifference map is homothetic to the origin. In Section 5 I consider several classes of non-homothetic preference orderings and examine their implications for the relationship between the cost-of-living index and the base indifference curve.

In Section 6 I consider a menu of preference orderings simple enough that they are suitable for empirical work and examine the cost-of-living index corresponding to each. Since computation of a cost-of-living index depends on estimation of an underlying system of demand equations, it is important to have some sense of what the possibilities are.

In Section 7 I follow Malmquist in defining a preference field quantity index. In Section 8 I show that use of a cost-of-living index to deflate the expenditure index does not yield this index unless the preference ordering is homothetic to the origin. This suggests that if one wants a quantity index, one must try to get it directly rather than by deflating expenditure by a price index.

The discussion in Section 9 of the use of price indexes in empirical demand analysis suggests that the common practice of deflating prices and income by a Laspeyres index has no theoretical standing. It also implies that deflating them by the cost-of-living index would be little better. Furthermore, since the cost-of-living index cannot be computed until the unknown parameters of the system of demand equations have been estimated, a procedure which required deflation of price and income by that index would be of little practical value.

The literature on the theory of the cost-of-living index is of uneven quality. The most lucid general treatments are those of Samuelson [1947], Wold and Jureen [1953], and Malmquist [1953]. A more recent survey is Afriat [1972]. Fisher and Shell [1968] give a concise statement of the general problem and a detailed discussion of a particular type of quality change.[1]

1. Preliminaries

An individual's tastes can be represented by a preference ordering defined over the commodity space. Let $X = (x_1,...,x_n)$ and $X' = (x'_1,...,x'_n)$ denote commodity bundles and write $X \, R \, X'$ for "X is at least as good as X'." The binary relation R is a preference ordering. If R satisfies the usual conditions (completeness, reflexivity, transitivity, convexity, continuity), then there exists a real valued function, U(X), which "represents" the preference ordering in the sense that X R X' if and only if $U(X) \geqq U(X')$. We call U(X) the "direct utility function" corresponding to the preference ordering R, and we sometimes write it as U(X;R) to emphasize the particular preference ordering, R, which the utility function represents. If U(X) is a direct utility function corresponding to the preference ordering, R, then any increasing monotonic transformation of U is also a direct utility function representing R.

The "ordinary demand functions" are found by maximizing the direct utility function subject to the budget constraint. Let $P = (p_1,...,p_n)$ denote the price vector, and μ total expenditure.[2] We denote the ordinary demand functions by $x_i = h^i(P,\mu)$ or $x_i = h^i(P,\mu;R)$, when we need to indicate explicitly the preference ordering from which the demand functions are derived. In vector form, $X = H(P,\mu)$ or $X = H(P,\mu;R)$.

We say that a system of demand functions exhibits "expenditure proportionality" if $h^i(P,\mu) = \gamma^i(P)\mu$ for all i. This is equivalent to requiring all income elasticities to be 1 or, equivalently, all income-consumption curves to be rays from the origin. We say that a preference ordering is "homothetic to the origin" if it can be represented by a utility function which is an increasing monotonic transformation of a function homogeneous of degree 1: $U(X) = T[g(X)]$, $g(\lambda X) = \lambda g(X)$. We often call the utility function itself "homothetic". It is well known that a system of demand functions exhibits expenditure proportionality if and only if it is generated by a preference ordering which is homothetic to the origin.

The "compensated" or "constant utility" demand functions are found by minimizing the cost of attaining a particular indifference curve. We select a direct utility function to represent the preference ordering and denote the level of utility corresponding to the indifference curve by s; we denote the compensated demand functions by $x_i = f^i(P,s)$ or $x_i = f^i(P,s;R)$, or, in vector form, $X = F(P,s)$.[3] The role of s, the level of utility associated with the indifference curve, requires some explanation. The compensated demand functions depend on the particular indifference curve chosen but not on the particular utility function selected to represent the preference ordering. It would be more accurate to denote the compensated demand functions by $x_i = f^i(P,s;R,U)$ to indicate that the interpretation of a particular numerical value of s depends on the direct utility function selected to represent R. Usually, this elaborate notation is unnecessary.

The direct utility function represents preferences by assigning a number to each X in the commodity space. One collection of goods is assigned a higher number than another if and only if the first is preferred to the second. In a similar manner we define an indirect utility function which represents preferences by assigning numbers to "price-expenditure situations", (P,μ): one price-expenditure situation is assigned a higher number than another if and only if the first is preferred to the second. The difficulty is that we have not yet defined what it means to say that one price-expenditure situation is preferred to another. The obvious meaning is that the best collection of goods available in the first situation is preferred to the best collection available in the second. Since $X = H(P,\mu;R)$ is the best collection of goods available at (P, μ), we say that (P, μ) is preferred to (P', μ') if and only if $X = H(P,\mu;R)$ is preferred to $X' = H(P',\mu';R)$. We can use the direct utility function, $U(X;R)$, to assign numbers to price-expenditure situations and thus to define the indirect utility function $\Psi(P,\mu;R)$:

$$\Psi(P,\mu;R) = U[H(P,\mu;R);R].$$

Thus, the indirect utility function is the maximum value of the direct utility function attainable in a particular price-expenditure situation:

$$\Psi(P,\mu;R) = \max U(X;R) \quad \text{subject to} \quad \Sigma p_k x_k \leq \mu.$$

The ordinary demand functions are related to the indirect utility function by

$$h^i(P,\mu) = -\frac{\dfrac{\partial \Psi(P,\mu)}{\partial p_i}}{\dfrac{\partial \Psi(P,\mu)}{\partial \mu}}$$

This result is often called Roy's theorem; since we use it repeatedly, we sketch a proof. (1) Differentiate $\Psi(P,\mu) = U[H(P,\mu)]$ with respect to p_i, replace U_k by $-\lambda p_k$ (from the first order conditions from maximization of the direct utility function), and replace $\Sigma p_k h_i^k$

by $-h^i$ (this follows from differentiating the budget constraint with respect to p_i). This yields $\Psi_i = \lambda h^i$. (2) Differentiate $\Psi(P,\mu) = U[H(P,\mu)]$ with respect to μ, replace U_k by $-\lambda p_k$, and recognize that $\Sigma\ p_k h^k_\mu = 1$. This yields $\Psi_\mu = -\lambda$. Hence, $-\Psi_i/\Psi_\mu = h^i$.

Finally, preferences can be represented by the "expenditure function", E(P,s), or E(P,s;R), analogous to the cost function in production theory.[4] In production theory the cost function shows the minimum cost of attaining a given level of output or, equivalently, a given isoquant; in consumer theory it shows the minimum expenditure required to attain a given level of utility or, strictly speaking, a given indifference curve. The expenditure function is related to the compensated demand functions by $E(P,s) = \Sigma\ p_k f^k(P,s)$.

Alternatively, the expenditure function may be derived from the indirect utility function by solving

$$s = \Psi(P,\mu)$$

for μ . Since

$$\frac{\partial E(P,s)}{\partial p_i} = - \frac{\dfrac{\partial \Psi[P,E(P,s)]}{\partial p_i}}{\dfrac{\partial \Psi[P,E(P,s)]}{\partial \mu}} = h^i[P,E(P,s)],$$

we have

$$\frac{\partial E(P,s)}{\partial p_i} = f^i(P,s).$$

2. Basic Theory

The cost-of-living index is the ratio of the minimum expenditures required to attain a particular indifference curve under two price regimes. We denote the cost-of-living index by $I(P^a,P^b,s,R)$:

$$I(P^a, P^b, s, R) = \frac{E(P^a, s, R)}{E(P^b, s, R)}$$

The notation emphasizes that the index depends not only on the two sets of prices, P^a and P^b, but also on an initial choice of an indifference map or preference ordering, R, and the choice of a base indifference curve, s, from that map. One set of prices is called "reference prices" and the other, "comparison prices". If comparison prices are twice the reference prices, the index is 2; if they are one-half the reference prices, the index is one-half. In our notation the comparison prices are the first n arguments of the index function, and reference prices, the next n arguments. Interchanging comparison and reference prices yields a new index which is the reciprocal of the original index:

$$I(P^a, P^b, s, R) = \frac{1}{I(P^b, P^a, s, R)} \, .$$

Either set of prices may be designated the reference set; a choice must be made, but it is a choice without substantive implications. Usually, we will denote comparison prices by P^a and reference prices by P^b. The reader is cautioned against calling the reference prices "base" prices; we reserve the term "base" to denote the indifference curve on which the index is predicated or the preference ordering to which it belongs.

Strictly speaking, the cost-of-living index depends only on the comparison prices, the reference prices, and the base indifference curve. It does not depend on the indifference map to which the base curve belongs. However, it is useful and realistic to imagine that the base indifference curve is selected by a two-stage procedure: first, a base map is chosen, and then a base curve is chosen from the map. Treating the base curve as part of a base map leads one to investigate the sensitivity of the index to the choice of the base curve from a particular map.

The logic of the cost-of-living index is best understood by interpreting it in the twin contexts of comparisons over time and comparisons over space. Features of the index which are easily overlooked in one context often stand out sharply in the other. This is particularly true of the role of the base preference ordering and base indifference curve.

For example, to construct a cost-of-living index to compare prices in Paris with those in Tokyo, we must specify the preference ordering on which the comparison is to be based. The Japanese government, considering how much to pay its diplomats in Paris, would presumably use Japanese tastes, while the French government would use French tastes. As is customary in discussing international price comparisons, we ignore differences in tastes within countries. But suppose the U.S. government wants to compare prices in Paris with those in Tokyo to decide on appropriate salary differentials for its diplomats. The comparison should be based on U.S. tastes. In principle, this is extremely important because it underscores the fact that the base preference ordering need not be one which is associated either with the reference prices or the comparison prices.

In intertemporal comparisons we often conceal the choice of a base preference ordering. To compare U.S. prices in 1970 with U.S. prices in 1969, it seems "obvious" and "natural" to use U.S. preferences. As a practical matter it seems likely that this would be appropriate in the majority of problems of this type although they would not be appropriate for the French government to use when deciding how much to increase the salary of its diplomats in Washington. The principal difficulty is that specifying "U.S. preferences" does not resolve the problem unless U.S. preferences are constant over time. Otherwise, it identifies a class of preference orderings from which the appropriate one still must be chosen. At first glance we have reduced the number of admissible preference orderings to two: U.S. preferences in 1969 and in 1970. But there is no reason why the comparison should not be based on U.S. preferences in 1971, 1958, or any other year. Fisher and Shell [1968] argue that current tastes provide a more appropriate indicator of the welfare effects of price changes than past tastes, but they are concerned only with the choice between 1969 and 1970 tastes and do not consider the possibility of basing the comparison on 1971 preferences.

The case of endogenous taste change is conceptually more difficult. If tastes change because of habit formation, as in Pollak [1970], then the appropriate base preference ordering may be a long-run pseudo preference ordering which generates the long-run demand functions rather than any particular short-run demand function with its implied dependence on the historic time path of consumption.[5] Endogenous taste change has received little systematic attention in economic theory; if tastes are endogenous, the validity of individual preferences as a touchstone of social welfare must be re-examined.

We conclude this section by enumerating the properties of the cost-of-living index which follow directly from its definition and the properties of the cost function:

$$I(P,P,s,R) = 1 \qquad \qquad \text{P1.}$$

That is, if the comparison prices are equal to the reference prices, the value of the index is 1.

$$I(\lambda P,P,s,R) = \lambda \qquad \qquad \text{P2.}$$

That is, if the comparison prices are proportional to the reference prices, then the value of the index is equal to the factor of proportionality.

$$I(P,\lambda P,s,R) = \frac{1}{\lambda} \qquad \qquad \text{P3.}$$

That is, if the reference prices are proportional to the comparison prices, then the value of the index is the reciprocal of the factor of proportionality.

$$I(\lambda P^a,\lambda P^b,s,R) = I(P^a,P^b,s,R). \qquad \qquad \text{P4.}$$

If the comparison prices and the reference prices are multiplied by a common factor, the value of the index is unchanged.

$$I(P^b,P^a,s,R) = 1/I(P^a,P^b,s,R). \qquad \qquad \text{P5.}$$

If the comparison and the reference prices are interchanged, then the new index is the reciprocal of the old.

$$\text{If } P^{a'} \geq P^a, \text{ then } I(P^{a'},P^b,s,R) \geq I(P^a,P^b,s,R). \qquad \qquad \text{P6.}$$

That is, if one set of comparison prices is higher than another, the index corresponding to the first is higher than that corresponding to the second. If the index is differentiable, we can express this monotonicity property as

$$\frac{\partial I}{\partial p_i^a}(P^a, P^b, s, R) \geq 0.$$

The strict inequality holds if all goods are consumed everywhere in a neighborhood of the initial price-expenditure situation. The property follows directly from the fact that an increase in any price cannot decrease the cost of attaining a particular indifference curve.

$$\min \left\{\frac{p_i^a}{p_i^b}\right\} \leq I(P^a, P^b, s, R) \leq \max \left\{\frac{p_i^a}{p_i^b}\right\}. \qquad \text{P7.}$$

The cost-of-living index for any base indifference curve lies between the smallest and the largest "price relative", p_i^a / p_i^b. To prove this we set μ^b so that $\Psi(P^b, \mu^b; R) = s$ and, therefore, $\mu^b = E(P^b, s; R)$. It suffices to show

$$\min \left\{\frac{p_i^a}{p_i^b}\right\} \leq \frac{E(P^a, s; R)}{\mu^b} \leq \max \left\{\frac{p_i^a}{p_i^b}\right\}$$

or, equivalently,

$$\mu^b \min \left\{\frac{p_i^a}{p_i^b}\right\} \leq E(P^a, s; R) \leq \mu^b \max \left\{\frac{p_i^a}{p_i^b}\right\}.$$

We now proceed with an overcompensation argument and an undercompensation argument. (1) If you give the individual $\mu^b \max \frac{p_i^a}{p_i^b}$, then he cannot be worse off than he was at P^b, μ^b because, regardless of what collection of goods he purchased at that price-expenditure situation, he can buy the same collection now with expenditure $\mu^b \max \frac{p_i^a}{p_i^b}$. In particular, this is true even if he consumed only one good, and that happened to be the good which experienced the largest price increase. (2) If you give an individual

$\mu^b \min \dfrac{p_i^a}{p_i^b}$, then he cannot be better off than at P^b, μ^b because the new feasible set lies

entirely within the old one except where they coincide at the vertex corresponding to the good whose price has experienced the smallest increase. These upper and lower bounds are important because they do not depend on knowing anything about preferences except that they satisfy the usual regularity conditions, and because they do not depend on knowing the quantities consumed in any price-expenditure situation.[6]

3. Laspeyres and Paasche Indexes

The theory of the cost-of-living index provides no criterion for choosing either the base map or the base curve on which the index is predicated. The upper and lower bounds on the cost-of-living index expressed in P7.,

$$\min \left\{ \frac{p_i^a}{p_i^b} \right\} \leq I(P^a, P^b, s, R) \leq \max \left\{ \frac{p_i^a}{p_i^b} \right\},$$

represent the best that can be done without additional assumptions. In this section we examine the cost-of-living index corresponding to two "indifference map-indifference curve" combinations which stand out as "natural" or "obvious" ones on which to base the index, namely, those which correspond to the reference situation and the comparison situation.

Consider an individual with preference ordering, R^a, who, facing prices P^a with expenditure μ^a chooses the basket of goods X^a : $X^a = H(P^a, \mu^a; R^a)$. Similarly, $X^b = H(P^b, \mu^b; R^b)$. The most suggestive interpretation is in terms of place-to-place comparisons. Suppose P^a and P^b denote prices in Paris and Tokyo, so R^a and R^b denote French and Japanese preferences. There are two "natural" or "obvious" indifference curves which stand out as candidates on which to base a cost-of-living index. If we are to use French tastes, it seems "natural" (although certainly not necessary) to consider the indifference curve attained by a Frenchman facing prices P^a with income μ^a. We define s^a by $s^a = \Psi(P^a, \mu^a; R^a)$. If we are to use Japanese tastes as our norm, it seems "natural" to consider the indifference curve attained by a Japanese facing prices P^b with expenditure μ^b: $s^b = \Psi(P^b,\mu^b;R^b)$. Thus, the two "natural" (s,R) combinations on which to base a living index are (s^a,R^a) and (s^b,R^b). In the case of intertemporal comparisons the situation is identical. However, if tastes do not change over time, then $R^a = R^b$, and the two "natural" base indifference curves belong to the same preference ordering.

We have identified two "natural" indexes, $I(P^a,P^b,s^a,R^a)$ and $I(P^a,P^b,s^b,R^b)$. There are other indexes which have some claim to being called "natural", and the primacy attributed to these two may reflect no more than the fact that we have interesting theorems about them. Two other "natural" indexes are $I(P^a,P^b,s^{a*},R^a)$ where $s^{a*} = \Psi(P^a,\mu^b;R^a)$ and $I(P^a,P^b,s^{b*},R^b)$ where $s^{b*} = \Psi(P^b,\mu^a;R^b)$. The first is based on the indifference curve which could be attained by an individual with the map of the comparison situation, facing comparison prices with the expenditure of the reference situation. The second is based on the curve attained by an individual with the map of the reference situation facing reference prices but with comparison expenditure.

The two indexes which we identified as natural are of special interest because we can establish better bounds for them than we could in the general case. To establish these bounds, we define a fixed weight index, $J(P^a,P^b,\theta)$,

$$J(P^a,P^b,\theta) = \frac{\Sigma\theta_k p_k^a}{\Sigma\theta_k p_k^b}$$

where $\theta = (\theta_1,...,\theta_n)$. The fixed weight index is a ratio of weighted sums of prices, but we could, without loss of generality, divide through by $\Sigma \theta_k$ and interpret the index as a ratio of weighted averages of prices. We shall interpret the weights as quantities of the goods in a market basket, so the index is the ratio of the cost of that market basket at prices P^a to its cost at prices P^b.

A fixed weight price index provides little useful information unless the weights are carefully chosen. Two obvious choices of weights are X^a and X^b. The fixed weight index with weights equal to X^b is called a "Laspeyres" index:

$$J(P^a,P^b,X^b) = \frac{\Sigma\, x_k^b p_k^a}{\Sigma\, x_k^b p_k^b}.$$

That is, the Laspeyres index is a fixed weight index with weights associated with the reference prices, P^b. We often write the Laspeyres index in the form

$$J(P^a,P^b,X^b) = \Sigma\, w_k^b \left(\frac{p_k^a}{p_k^b}\right)$$

where $w_k^b = \dfrac{x_k^b p_k^b}{\mu^b}$. That is, the Laspeyres index is a weighted average of the "price relatives", p_i^a/p_i^b, where the weights are the expenditure weights of the reference situation. To show the equivalence of these two forms, we write

$$J(P^a,P^b,X^b) = \frac{\Sigma\, x_k^b p_k^a}{\mu^b} = \Sigma\, \frac{x_k^b}{\mu^b}\, p_k^a$$

$$= \Sigma\, \frac{x_k^b p_k^b}{\mu^b}\, \frac{p_k^a}{p_k^b} = \Sigma\, w_k^b \left(\frac{p_k^a}{p_k^b}\right).$$

We use the fact that X^b is the market basket purchased by an individual with preferences R^b with expenditure μ^b at prices P^b to establish an upper bound on $I(P^a,P^b,s^b,R^b)$.

R.A. Pollak

Theorem:

$$\min \left\{ \frac{p_i^a}{p_i^b} \right\} \leq I(P^a, P^b, s^b, R^b) \leq J(P^a, P^b, X^b) \ .$$

The lower bound is the one asserted in P7., but the upper bound is an improvement since

$$J(P^a, P^b, X^b) = \Sigma \ w_k^b \left(\frac{p_k^a}{p_k^b} \right) \leq \max \left\{ \frac{p_k^a}{p_k^b} \right\} .$$

since the w's are non-negative numbers which sum to 1. That is, the Laspeyres index is an upper bound on the cost-of-living index based on the indifference curve attained in the reference situation.

Proof: To show that $J(P^a, P^b, X^b)$ is an upper bound on $I(P^a, P^b, s^b, R^b)$ we write the latter as

$$\frac{E(P^a, s^b, R^b)}{E(P^b, s^b, R^b)} = \frac{E(P^a, s^b, R^b)}{\mu^b}$$

and the former as

$$\frac{\Sigma \ x_k^b p_k^a}{\Sigma \ x_k^b p_k^b} = \frac{\Sigma \ x_k^b p_k^a}{\mu^b} .$$

It suffices to show that

$$E(P^a, s^b, R^b) \leq \Sigma \ x_k^b p_k^a \ .$$

But this follows directly from the fact that the minimum cost of attaining s^b at prices P^a cannot be greater than the cost of X^b.

The Paasche index is the fixed weight index with weights equal to the market basket purchased at the comparison prices, P^a, with expenditure μ^a:

$$J(P^a, P^b, X^a) = \frac{\Sigma \, x_k^a p_k^a}{\Sigma \, x_k^a p_k^b} \, .$$

It is the ratio of the cost of buying the market basket X^a at prices P^a to its cost at P^b. The Paasche index is a lower bound on $I(P^a, P^b, s^a, R^a)$, the cost-of-living index corresponding to (s^a, R^a).

Theorem:

$$J(P^a, P^b, X^a) \le I(P^a, P^b, s^a, R^a) \le \max \left\{ \frac{p_i^a}{p_i^b} \right\} .$$

Proof: The upper bound is the one established in P7. To establish the lower bound, we must show that

$$\frac{\Sigma x_k^a p_k^a}{\Sigma \, x_k^a p_k^b} \le \frac{E(P^a, s^a, R^a)}{E(P^b, s^a, R^a)} \, .$$

Since the numerators are equal, it suffices to show

$$\frac{1}{\Sigma \, x_k^a p_k^b} \le \frac{1}{E(P^b, s^a, R^a)}$$

or, equivalently,

$$E(P^b, s^a, R^a) \le \Sigma \, x_k^a p_k^b \, .$$

But this follows immediately since the minimum expenditure required to attain (s^a, R^a)

at prices P^b cannot exceed the cost of X^a at these prices.

To summarize: the Laspeyres index is a fixed weight index with weights corresponding to the market basket purchased in the reference situation. It is an upper bound on the cost-of-living index corresponding to the preference ordering and indifference curve attained in the reference situation. The Paasche index is a fixed weight index with weights corresponding to the market basket purchased in the comparison situation. It is a lower bound on the cost-of-living index corresponding to the preference ordering and indifference curve attained in the comparison situation. It is not true that the cost-of-living index lies between the Paasche and Laspeyres indexes. Instead, we have a lower bound on one cost-of-living index and an upper bound on another.

4. When the Index is Equal to Its Bounds

In Section 3 we established two important bounding theorems for the cost-of-living index:

$$\min \left\{ \frac{p_i^a}{p_i^b} \right\} \leq I(P^a, P^b, s^b, R^b) \leq J(P^a, P^b, X^b)$$

and

$$J(P^a, P^b, X^a) \leq I(P^a, P^b, s^a, R^a) \leq \max \left\{ \frac{p_i^a}{p_i^b} \right\}.$$

In this section we investigate the preference orderings for which the cost-of-living index coincides with one or the other of its bounds.

4.1 When the cost-of-living index is equal to the Laspeyres or Paasche bounds

It is well known that if the preference ordering is represented by a "fixed coefficient" direct utility function

$$U(X) = \min\left\{\frac{x_i}{a_i}\right\}$$

then the cost-of-living index $I(P^a,P^b,s^b,R)$ is equal to the Laspeyres index $J(P^a,P^b,X^b)$. To show this we make use of the fact that the expenditure minimizing quantity of good i for attaining a level of utility s^b is given by $x_i^b = a_i s^b$. Hence, the cost of attaining s^b at prices P^b is given by $\Sigma\ x_k^b p_k^b = \Sigma\ a_k s^b p_k^b$ while the minimum expenditure required to attain s^b at prices P^a is given by $\Sigma\ x_k^b p_k^a = \Sigma\ a_k s^b p_k^a$. Hence, the Laspeyres index $J(P^a,P^b,X^b)$ coincides with the cost-of-living index $I(P^a,P^b,s^b,R)$.

The homothetic fixed coefficient case is not the only one in which $I(P^a,P^b,s^b,R) = J(P^a,P^b,X^b)$. Any preference ordering which does not permit substitution along its indifference curves implies a cost-of-living index, $I(P^a,P^b,s^b,R)$, which coincides with the Laspeyres index, $J(P^a,P^b,X^b)$. There is no need for the indifference map to be homothetic.

Theorem: The cost-of-living index coincides with the appropriate Laspeyres and Paasche bounds if and only if the preference ordering can be represented by a direct utility function of the generalized fixed coefficient form

$$U(X) = \min\left\{g^i(x_i)\right\} \qquad g^{i\,\prime}(x^i) > 0.$$

Proof: If the preference ordering is of the generalized fixed coefficient form, it is easily verified that the cost-of-living index coincides with the appropriate Laspeyres and Paasche bounds.

If the cost-of-living index coincides with its Laspeyres bound for all s^b, then

$$\Psi(P^b,m) = \Psi[P^a,\Sigma\ h^k(P^b,m)p_k^a]$$

for all m. Differentiating with respect to p_i^b and m, we find the ordinary demand functions:

$$-h^i(P^b,m) = \frac{\Sigma \ p_k^a \ \dfrac{\partial \ h^k}{\partial \ p_i^b}}{\Sigma \ p_k^a \ \dfrac{\partial \ h^k}{\partial \ m}} .$$

Hence,

$$\Sigma \ p_k^a \left[\frac{\partial \ h^k}{\partial \ p_i^b} + h^i \ \frac{\partial \ h^k}{\partial \ m} \right] = 0$$

or, differentiating with respect to p_j^a,

$$\frac{\partial \ h^j}{\partial \ p_i^b} + h^i \frac{\partial \ h^j}{\partial \ m} = 0 .$$

This, of course, implies that the substitution effects are zero. We next show that the demand functions $h^2,...,h^n$ can each be written as functions of h^1:

$$h^i(P,m) = \delta^i[h^1(P,m)] .$$

We do this by showing that the ratio of the partial derivatives is equal

$$\frac{\dfrac{\partial \ h^j}{\partial \ p_i}}{\dfrac{\partial \ h^j}{\partial \ m}} = \frac{\dfrac{\partial \ h^1}{\partial \ p_i}}{\dfrac{\partial \ h^1}{\partial \ m}} \qquad \text{for all i,j .}$$

Equality of the partial derivatives follows from our characterization of the substitution effect; indeed, the common value of the ratios is h^1. Substituting these demand functions into the direct utility function yields the indirect utility function

$$s = \Psi(P,\mu) = U[h^1(P,\mu),\delta^2(h^1),...,\delta^n(h^1)] = \delta[h^1(P,\mu)] .$$

This implies $x_1 = f^1(P,s) = f^1(s)$, and, hence, $x_i = f^i(P,s) = f^i(s)$. This implies that the direct utility function is of the generalized fixed coefficient form where g^i is the inverse of f^i.

The Laspeyres (Paasche) index may coincide with the cost-of-living index to which it is the upper (lower) bound for a particular value of s, say s^*, but not for all s. This occurs if the indifference curve corresponding to s^* is of the fixed coefficient form. In another context Marjorie McElroy [1969] has provided an interesting example of such an indifference map; she constructed it by allowing the "necessary basket" of a linear expenditure system to coincide with the "bliss point" of an additive quadratic.[7] At the critical point, the Laspeyres and Paasche indexes coincide with the corresponding cost-of-living index.

4.2 When the cost-of-living index is equal to the "other bounds"

In Section 4.1 we showed that the cost-of-living index coincides with the appropriate Laspeyres or Paasche bounds if and only if the preference ordering is of the generalized fixed coefficient form. We now examine the two forgotten bounds: $\left\{\dfrac{p^a_i}{p^b_i}\right\}$, the lower bound of $I(P^a,P^b,s^b,R^b)$ and $\max\left\{\dfrac{p^a_i}{p^b_i}\right\}$, the upper bound of $I(P^a,P^b,s^a,R^a)$. For what preference orderings do these cost-of-living indexes coincide with their bounds?

If the preference ordering can be represented by a linear direct utility function

$$U(X) = \Sigma\, a_k x_k$$

then the indifference curves are parallel lines, and all goods are "perfect substitutes". The minimum cost of attaining the indifference curve s is given by

$$E(P,s,R) = \min\left\{\frac{p_i}{a_i}s\right\} = s \min\left\{\frac{p_i}{a_i}\right\}.$$

Hence, the cost-of-living index is

$$I(P^a,P^b,s,R) = \frac{\min\left\{\frac{p_i^a}{a_i}\right\}}{\min\left\{\frac{p_i^b}{a_i}\right\}}.$$

The ordinary demand functions corresponding to this utility are not single valued. If

$$\frac{p_1}{a_1} = \frac{p_2}{a_2} = \ldots = \frac{p_n}{a_n},$$ then the budget line coincides with the indifference curve cor-

responding to $s = \dfrac{\mu a_i}{p_i}$, and the consumer is indifferent among all commodity bundles

which exhaust his expenditure. For any other configuration of relative prices some goods will not be consumed.

Suppose that an individual's preferences are represented by a linear utility function and that when facing prices P^b with expenditure μ^b, he consumes all goods in positive quantities. Then

$$\frac{p_1^b}{a_1} = \frac{p_2^b}{a_2} = \ldots = \frac{p_n^b}{a_n} = r,$$ so the cost of living index is given by

$$I(P^a,P^b,s^b,R) = \frac{1}{r}\min\left\{\frac{p_i^a}{a_i}\right\} = \min\left\{\frac{p_i^a}{a_i}\right\}\frac{1}{r} = \min\left\{\frac{p_i^a}{a_i}\frac{a_i}{p_i^b}\right\} = \min\left\{\frac{p_i^a}{p_i^b}\right\}.$$

A similar result holds for the generalized linear direct utility function, $s = U(X)$, defined implicitly by

$$\Sigma \, \alpha^k(s) \, x_k = w(s),$$

where $\alpha^1, \ldots, \alpha^n$ and w are functions of s. The cost-of-living is given by

$$I(P^a, P^b, s, R) = \frac{\min \left\{ \dfrac{p_i^a}{\alpha^i(s)} \right\}}{\min \left\{ \dfrac{p_i^b}{\alpha^i(s)} \right\}}.$$

The indifference curves are linear, but they are not parallel; of course, they cannot intersect in the positive orthant, but there is no reason to rule out intersections of extensions of the indifference curves outside the commodity space. If all goods are consumed at (P^b, μ^b),

then it is easy to show that $I(P^a, P^b, s^b, R) = \min \left\{ \dfrac{p_i^a}{p_i^b} \right\}$. This is what one would expect

since the relevant characteristic in our previous example is that the indifference curves are linear; whether or not they are parallel is irrelevant.

We emphasize that it was necessary to assume that all goods are consumed at (P^b, μ^b). Suppose $p_i^a = p_i^b$, $i = 2, \ldots, n$ and $p_1^a < p_1^b$. If x_1 were not consumed at (P^b, μ^b) because it was too expensive, then a small decrease in its price, all other prices remaining constant, will not affect the cost-of-living index; x_1 will still be too expensive and will not be con-

sumed. Hence, the value of the index would be 1, not $\min \left\{ \dfrac{p_i^a}{p_i^b} \right\}$.

Theorem: The cost-of-living index coincides with the appropriate "other bounds",

$\min \left\{ \dfrac{p_i^a}{p_i^b} \right\}$ or $\max \left\{ \dfrac{p_i^a}{p_i^b} \right\}$, if and only if the preference ordering can be represented by a generalized linear utility function

$$\Sigma \, \alpha^k(s) \, x_k \, = \, w(s) \,,$$

and all goods are consumed in positive quantities in the base situation.

Proof: We have already proved that if the utility function is of this form, then $I(P^a, P^b, s^b, R)$

coincides with $\min \left\{ \dfrac{p_i^a}{p_i^b} \right\}.$

We now show that the preference ordering corresponding to the generalized linear direct utility function is the only one for which

$$I(P^a, P^b, s^b, R) \, = \, \min \left\{ \dfrac{p_i^a}{p_i^b} \right\}$$

provided that all goods are consumed at prices P^b with expenditure μ^b. We cannot vary the p_i^b's because such variations may invalidate the hypothesis that all goods are consumed. We can, however, vary the p_i^a's. If $\dfrac{p_1^a}{p_1^b} < \dfrac{p_i^a}{p_i^b}$ for all $i \neq 1$, then in a neighborhood of P^a the index depends only on p_1^a and is independent of $p_2^a, \ldots, p_n^a$. Hence, in that neighborhood $E(P^a, s^b)$ depends only on p_1^a and s. Since the compensated demand functions are the derivatives of the expenditure function,

$$f^i(P^a, s^b) \, = \, \dfrac{\partial \, E(P^a, s^b)}{\partial \, p_i^a} \, = \, 0 \qquad i \neq 1.$$

That is, only good 1 is consumed. This is the case in every price-expenditure situation unless there are "ties", and this implies linear indifference curves.

We now turn briefly to the index $I(P^a, P^b, s^a, R^a)$ and its upper bound $\max \left\{ \dfrac{p_i^a}{p_i^b} \right\}$. If the preference ordering can be represented by the generalized linear direct utility function, then

$$I(P^a, P^b, s^a, R^a) = \frac{\min \left\{ \dfrac{p_i^a}{\alpha_k(s^a)} \right\}}{\min \left\{ \dfrac{p_i^b}{\alpha_k(s^a)} \right\}}.$$

If all goods are consumed at (P^a, μ^a), then

$$\frac{p_1^a}{\alpha_1(s^a)} = \frac{p_2^a}{\alpha_2(s^a)} = \ldots = \frac{p_n^a}{\alpha_n(s^a)} = r$$

so that the cost-of-living index becomes

$$I(P^a, P^b, s^a, R) = \frac{r}{\min \left\{ \dfrac{p_i^b}{\alpha_i(s^a)} \right\}} = \frac{1}{\min \left\{ \dfrac{p_i^b}{\alpha_i(s^a)} \dfrac{1}{r} \right\}} = \frac{1}{\min \left\{ \dfrac{p_i^b}{p_i^a} \right\}}$$

$$= \max \left\{ \frac{p_i^a}{p_i^b} \right\}.$$

This result is precisely analogous to that obtained for $I(P^a, P^b, s^b, R)$, as indeed it must be.

The importance of the existence of preference orderings for which the cost-of-living index actually attains its "other bounds" lies in its immediate implication that these bounds are "best bounds". That is, if anyone claims to have found better bounds for the cost-of-living index, we can always find an admissible preference ordering whose cost-of-living index lies outside the proposed bounds. Although our "other bounds" may not seem as satisfying or as useful as the Laspeyres and Paasche bounds, our demonstration that they correspond to the generalized linear utility function shows that it is not our lack of ingenuity but the inherent logic of the situation which prevents us from finding better ones.

5. The Base Indifference Curve

In this section I examine how the choice of the base indifference curve affects the cost-of-living index.

5.1 Expenditure proportionality and homothetic indifference maps

If the preference ordering is homothetic to the origin, then the implied cost-of-living index is independent of the particular indifference curve chosen as a base. That is, if R is homothetic to the origin, $I(P^a, P^b, s, R)$ is independent of s. To prove this, we use the fact that if a preference ordering is homothetic to the origin, then it can be represented by a direct utility function homogeneous of degree 1; the implied demand functions exhibit expenditure proportionality; and the indirect utility function can be written in the form

$$s = \Psi(P, \mu) = \phi(P)\mu$$

where $\phi(P)$ is homogeneous of degree -1. Hence, the expenditure function is given by

$$\mu = E(P, s) = \frac{s}{\phi(P)}$$

and the cost-of-living index by

$$I(P^a, P^b, s, R) = \frac{E(P^a, s, R)}{E(P^b, s, R)} = \frac{\phi(P^b)}{\phi(P^a)}$$

which is independent of s.

The converse of this result also holds: the cost-of-living index is independent of the base indifference curve if and only if the preference ordering is homothetic to the origin. Instead of showing this directly, we first introduce what appears to be a roundabout way of specifying the base indifference curve from a given preference ordering R. We specify the base indifference curve as the one corresponding to a base level of expenditure, m, at reference prices, P^b. The level of utility corresponding to the base indifference curve is given by s = $\Psi(P^b,m)$. It is often convenient to write the index as a function of m rather than s. We write $I^*(P^a,P^b,m,R) = I(P^a,P^b,\Psi(P^b,m),R)$. The index $I(P^a,P^b,s,R)$ is independent of s if and only if $I^*(P^a,P^b,m,R)$ is independent of m.

The specification of the index as a function of m rather than s is more than a useful mathematical trick. In practice, it is a sensible, convenient and commonly-used method of specifying the base indifference curve. That is, the base indifference curve is specified to be the indifference curve from a given base indifference map attainable by an individual with a particular expenditure at base period prices. In fact, we cannot interpret the index $I(P^a,P^b,s,R)$ without additional information which enables us to attach some meaning to the numerical value of s. To do this we need either the direct utility function, the indirect utility function, or the expenditure function. The cost-of-living index, $I^* = I^*(P^a,P^b,m,R)$, is defined implicitly by

$$\Psi(P^b,m) = \Psi(P^a,mI^*) .$$

Differentiating with respect to p_i^b and m and making use of the assumption that

$$\frac{\partial I^*(P^a,P^b,m,R)}{\partial m} = 0 \text{ yields}$$

$$\frac{\partial \Psi(P^b,m)}{\partial p_i^b} = \frac{\partial \Psi(P^b,mI^*)m}{\partial \mu} \frac{\partial I^*}{\partial p_i^b}$$

$$\frac{\partial \Psi(P^b,m)}{\partial m} = \frac{\partial \Psi(P^a,mI^*)I^*}{\partial \mu}$$

Hence,

$$h^i(P^b,m) = -\frac{\dfrac{\partial \Psi(P^b,m)}{\partial p_i^b}}{\dfrac{\partial \Psi(P^b,m)}{\partial m}} = -\left(\frac{1}{I^*}\frac{\partial I^*}{\partial p_i^b}\right)m \ .$$

Since the factor in parenthesis is independent of m, the demand functions exhibit expenditure proportionality, and, therefore, the preference ordering is homothetic to the origin.

We have just proved:

Theorem: The cost-of-living index is independent of the base indifference curve if and only if the preference ordering is homothetic to the origin.

This implies:

Theorem: If the preference ordering is homothetic to the origin, then

$$J(P^a,P^b,X^a) \leq I(P^a,P^b,s,R) \leq J(P^a,P^b,X^b) \ .$$

That is, if the preference ordering is homothetic, then the cost-of-living index lies between its Paasche and Laspeyres bounds. This follows immediately from our previous theorem which implies that the cost-of-living index is independent of the base indifference curve.

These results are important not because we believe that peoples' indifference maps are homothetic but because we believe they are not. Our theorem, therefore, implies that the cost-of-living index depends on the choice of the base level of expenditure. We now investigate the ways in which the preference ordering determines the relationship between the cost-of-living index and the base level of expenditure.

5.2 Demand functions locally linear in expenditure

We say that a system of demand functions is locally linear in expenditure if

$$h^i(P,\mu) = \chi_i(P) + \delta_i(P)\mu, \qquad i = 1,...,n.$$

These demand functions are of substantially more empirical interest than those exhibiting expenditure proportionality. We now examine the form of the cost-of-living index implied by the preference ordering corresponding to these demand functions. W.M Gorman [1961] has shown that a system of demand functions is locally linear in expenditure if and only if its indirect utility function can be written in the form

$$s = \Psi(P,\mu) = \frac{\mu - f(P)}{g(P)},$$

where $f(P)$ and $g(P)$ are functions homogeneous of degree 1. The implied expenditure function is given by

$$\mu = E(P,s) = f(P) + g(P)s$$

and the cost-of-living index by

$$I^*(P^a,P^b,m,R) = \frac{1}{m} \left[f(P^a) + g(P^a) \left(\frac{m - f(P^b)}{g(P^b)} \right) \right]$$

$$= \alpha(P^a,P^b) + \frac{1}{m} \beta(P^a,P^b)$$

where

$$\alpha(P^a,P^b) = \frac{g(P^a)}{g(P^b)}$$

$$\beta(P^a,P^b) = f(P^a) - f(P^b)\frac{g(P^a)}{g(P^b)}.$$

That is, if the demand functions are locally linear in expenditure, then the cost-of-living index is linear in the reciprocal of base expenditure. If $f(P) = 0$, then the indifference map is homothetic to the origin, and the index is independent of m. As m approaches ∞, the cost-of-living index approaches a finite limit, and the influence of m becomes negligible. However, both of these assertions must be viewed cautiously since $f(P) = 0$ and m $\to \infty$ are inadmissible cases for certain g's. The quadratic case (c = 2) of Section 6.3 provides an illustration of both possibilities.

The converse of our characterization of the cost-of-living index also holds:

Theorem: The cost-of-living index depends linearly on the reciprocal of base expenditure if and only if the demand functions are locally linear in expenditure.

Proof: If $I = \alpha + \dfrac{\beta}{m}$, then the index is implicitly defined by

$$\Psi[P^b,m] = \Psi(P^a,mI) = \Psi(P^a,\alpha m + \beta) .$$

Differentiating with respect to p_i^b and m and recognizing that the demand functions are the negative of the ratios of these derivatives yields

$$h^i(P^b,m) = \frac{\alpha_i}{\alpha} m + \frac{\beta_i}{\alpha} .$$

5.3 Cost-of-living index linear in base expenditure

Theorem: If the cost-of-living index is linear in base expenditure, $I(P^a,P^b,m,R) = \alpha(P^a,P^b) + \beta(P^a,P^b)m$, then the demand functions are of the "Tornquist" form

$$h^i(P,\mu) = \frac{\alpha_i m + \beta_i m^2}{\alpha + 2\beta m} .$$

Proof: The index is implicitly defined by

$$\Psi[P^b,m] = \Psi(P^a,mI) = \Psi[P^a,\alpha m + \beta m^2] .$$

It is easily verified that the implied demand functions are of the Tornquist form.

This is an interesting result since Wold and Jureen [1953, p.3] suggest that demand functions of this form are not unreasonable. These results hold only for a limited range of values of m. The cost-of-living index cannot be linear for all m unless $\beta = 0$, in which case we are back to expenditure proportionality and homothetic indifference maps. If $\beta \neq 0$, the linear cost-of-living index would soon violate either the upper or lower bounds

$$\min \left\{ \frac{p_i^a}{p_i^b} \right\} \leq I^*(P^a, P^b, m, R) \leq \max \left\{ \frac{p_i^a}{p_i^b} \right\} \ .$$

6. Specific Preference Orderings

In this section I examine the cost-of-living-index formulae which correspond to specific preference orderings. In 6.1 I consider those corresponding to demand functions which exhibit expenditure proportionality and are generated by an additive direct utility function. In 6.2 I examine the cost-of-living indexes corresponding to demand functions which are locally linear in expenditure and are generated by an additive direct utility function. In 6.3 I consider the quadratic direct utility function. Irving Fisher's "ideal index", the geometric mean of the Laspeyres and the Paasche, is equal to the cost-of-living index if and only if the direct utility function is a homogeneous quadratic.

6.1 Additive utility functions and expenditure proportionality[8]

If an individual's utility function is additive and his demand functions exhibit expenditure proportionality, then his utility function belongs to the "Bergson family"

$$U(X) = \Sigma \, a_k \log x_k \qquad a_i > 0 \qquad \Sigma \, a_k = 1 \quad {}^9 \qquad (6.1.1)$$

$$U(X) = -\Sigma a_k x_k^c \qquad a_i > 0 \qquad c < 0 \qquad (6.1.2)$$

$$U(X) = \Sigma \, a_k x_k^c \qquad a_i > 0 \qquad 0 < c < 1 \qquad (6.1.3)$$

$$U(X) = \min \left\{ \frac{x_k}{a_k} \right\} \qquad a_i > 0 \qquad (6.1.4)$$

The demand functions corresponding to (6.1.1) are of the form

$$h^i(P,\mu) = \frac{a_i \mu}{p_i} \qquad (6.1.5)$$

while those corresponding to (6.1.2) and (6.1.3) are of the form

$$h^i(P,\mu) = \frac{\left(\dfrac{p_i}{a_i}\right)^{\frac{1}{c-1}} \mu}{\Sigma \, p_k \left(\dfrac{p_k}{a_k}\right)^{\frac{1}{c-1}}}. \qquad (6.1.6)$$

The demand functions corresponding to the fixed coefficient case (6.1.4) are

$$h^i(P,\mu) = \frac{a_i \mu}{\Sigma \, a_k p_k}. \qquad (6.1.7)$$

The indifference maps of these utility functions are identical with the isoquant maps of the C.E.S. production functions.

Since the demand functions exhibit expenditure proportionality, the corresponding indirect utility functions are of the form

$$\Psi(P,\mu) = \frac{\mu}{g(P)} \tag{6.1.8}$$

where $g(P)$ is homogeneous of degree 1. For the Cobb-Douglas case (6.1.1)

$$g(P) = \pi \, p_k^{a_k}. \tag{6.1.9}$$

In the C.E.S. cases, (6.1.2) and (6.1.3),

$$g(P) = [\Sigma \, a_k^{-\frac{1}{c-1}} \, p_k^{\frac{1}{c-1}}]^{\frac{c-1}{c}} \tag{6.1.10}$$

and in the fixed coefficient case (6.1.4)

$$g(P) = \Sigma \, a_k p_k \, . \tag{6.1.11}$$

The indirect utility function (6.1.8) implies an expenditure function of the form

$$\mu = E(P,s) = g(P)s \tag{6.1.12}$$

so the cost-of-living index is given by

$$I(P^a,P^b,s,R) = \frac{E(P^a,s,R)}{E(P^b,s,R)} = \frac{g(P^a)}{g(P^b)}. \tag{6.1.13}$$

As we showed in Section 5, expenditure proportionality is a necessary and sufficient condition for independence of the base indifference curve.

Two cases deserve special mention.

Theorem: The cost-of-living index is a geometric mean of the price relatives with weight independent of s

$$I(P^a, P^b, s, R) = \pi \left(\frac{p_k^a}{p_k^b} \right)^{a_k} \qquad (6.1.14)$$

if and only if the utility function is of the Cobb-Douglas form (6.1.1).

Proof: It is easy to verify that if the utility function is of the form (6.1.1), then the cost-of-living index is given by (6.1.14), and the weights $(a_1, \ldots, a_n)$ are the budget shares: $a_i = p_i h^i(P, \mu)$. To prove the converse, write

$$\Psi(P^b, m) = \Psi \left[P^a, m\pi \left(\frac{p_k^a}{p_k^b} \right)^{a_k} \right]$$

differentiate with respect to p_i^b and m, and verify that the implied demand functions are given by $h^i(P^b, m) = a_i m / p_i^b$.

One way in which the Cobb-Douglas case is special is that the cost-of-living index is a function of the price relatives p_i^a / p_i^b. The class of preference orderings with cost-of-living indexes of this type is a generalization of the Cobb-Douglas class.

Theorem: If the cost-of-living index is a function of price relatives,

$$I(P^a, P^b, s, R) = \hat{I} \left(\frac{P^a}{P^b}, s, R \right),$$

then the index is a geometric mean

$$I(P^a, P^b, s, R) = \pi \left(\frac{p_k^a}{p_k^b} \right)^{a^k(s)},$$

and the underlying preference ordering is a generalized Cobb-Douglas whose indirect utility function, $\Psi(P, \mu)$, is defined implicitly by

$$\Sigma \, \beta^k(s) \log p_k \, - \, \Sigma \, \beta^k(s) \log \mu \, = \, 1$$

where $\quad a^i(s)$ is defined by $a^i(s) \, = \, \beta^i(s)/\Sigma \, \beta^k(s)$. [10]

where $a^i(s)$ is defined by $a^i(s) \, = \, \beta^i(s)/\Sigma \, \beta^k(s)$.[10]

It is sometimes thought that constructing a cost-of-living index is a matter of finding an appropriate way to combine the price relatives. This theorem shows that such a view is incorrect and that, except in the generalized Cobb-Douglas case, the comparison and reference prices do not enter the cost-of-living index in ratio form. Furthermore, the only admissible cost-of-living index based on price relatives is their geometric mean.

Proof: If the indirect utility function $\Psi(P,\mu)$ is implicitly defined by

$$\Sigma \, \beta^k(s) \log p_k \, - \, \log \mu \, \Sigma \, \beta^k(s) \, = \, 1 \, .$$

We solve for $\log \mu$ as a function of s:

$$\log \mu \, = \frac{\Sigma \, \beta^k(s) \log p_k \, - \, 1}{\Sigma \, \beta^k(s)}$$

The logarithm of the cost-of-living index is given by

$$\log I(P^a,P^b,s,R) \, = \, \log \frac{\mu^a}{\mu^b} \, = \, \log \mu^a \, - \, \log \mu^b \, =$$

$$= \frac{\Sigma \, \beta^k(s) \log p_k^a \, - \, \Sigma \, \beta^k(s) \log p_k^b}{\Sigma \, \beta^k(s)} \, .$$

Let $\alpha^i(s) \, = \, \beta^i(s)/\Sigma \, \beta^k(s)$. Then

$$\log I(P^a,P^b,s,R) \, = \, \Sigma \, a^k(s) \log \frac{p_k^a}{p_k^b} = \, \log \, \pi \left(\frac{p_k^a}{p_k^b} \right)^{a^k(s)}$$

so

$$I(P^a, P^b, s, R) = \pi \left(\frac{p_k^a}{p_k^b} \right) a^k(s) .$$

The demand functions corresponding to this utility function are given by $h^i(P, \mu) = a^i(s)\mu/p_i$. This can be verified by differentiating

$$\Sigma \, \beta^k(s) \log p_k - \log \mu \, \Sigma \, \beta^k(s) = 1$$

with respect to p_i and and solving for $\dfrac{\partial \Psi}{\partial p_i}$ and $\dfrac{\partial \Psi}{\partial \mu}$.

We now show that this is the only preference ordering to yield a cost-of-living index which depends on price relatives. If

$$I(P^a, P^b, s, R) = \hat{I} \left(\frac{P^a}{P^b}, s, R \right) ,$$

then

$$\frac{E(p_1^a, \ldots, \lambda p_i^a, \ldots, p_n^a, s)}{E(p_1^b, \ldots, \lambda p_i^b, \ldots, p_n^b, s)} = I(P^a, P^b, s, R) .$$

Differentiating with respect to λ and setting $\lambda = 1$ yields

$$p_i^a \frac{\partial E(P^a, s)}{\partial p_i^a} = I(P^a, P^b, s, R) p_i^b \frac{\partial E(P^b, s)}{\partial p_i^b}$$

so

$$\frac{f^i(P^a,s)p_i^a}{f^i(P^b,s)p_i^b} = I(P^a,P^b,s,R) = \frac{E(P^a,s)}{E(P^b,s)}$$

or, equivalently,

$$\frac{f^i(P^a,s)p_i^a}{E(P^a,s)} = \frac{f^i(P^b,s)p_i^b}{E(P^b,s)}.$$

That is, along an indifference curve the expenditure weight of each good is independent of prices: we denote the expenditure weight of good i by $a^i(s)$. But any system of demand functions of this form can be generated by the generalized Cobb-Douglas.

We showed in Section 4 that if the demand functions are generated by a homogeneous fixed coefficient utility function, then the Laspeyres and Paasche indexes coincide with each other, are independent of the base indifference curve, and coincide with the cost-of-living index. For completeness we restate that result here:

Theorem: If the direct utility function is of the homogeneous fixed coefficient form (6.1.4), then the cost-of-living index $I(P^a,P^b,s,R)$ coincides with the Laspeyres and Paasche indexes:

$$I(P^a,P^b,s,R) = \frac{\Sigma\, a_k p_k^a}{\Sigma\, a_k p_k^b} = \frac{\Sigma\, h^k(P,\mu)p_k^a}{\Sigma\, h^k(P,\mu)p_k^b} = J(P^a,P^b,X^b) = J(P^a,P^b,X^a) \qquad (6.1.15)$$

Since the Laspeyres index can be written as

$$J(P^a,P^b,X^b) = \Sigma\, w_k^b \left(\frac{p_k^a}{p_k^b}\right),$$

it might be thought that the homogeneous fixed coefficient case is one in which the cost-of-living index depends on price relatives. But this is not the case because the weights themselves depend on reference prices and not on the price relatives:

$$w_k^b = \frac{p_k^b h^k(P^b, \mu^b)}{\mu^b} = \frac{a_i p_i^b}{\Sigma a_k p_k^b}.$$

6.2 Additive utility functions and linear Engel curves

In this section I examine the cost-of-living index corresponding to demand functions which are locally linear in income and which are generated by additive direct utility functions. In Pollak [1971] I showed that the utility functions

$$U(X) = \Sigma a_k \log (x_k - b_k) \qquad a_i > 0, (x_i - b_i) > 0, \Sigma a_k = 1. \qquad (6.2.1)$$

$$U(X) = -\Sigma a_k(x_k - b_k)^c \qquad c < 0, (x_i - b_i) > 0. \qquad (6.2.2)$$

$$U(X) = \Sigma a_k(x_k - b_k)^c \qquad 0 < c < 1, a_i > 0, (x_i - b_i) > 0. \qquad (6.2.3)$$

$$U(X) = -\Sigma a_k (b_k - x_k)^c \qquad c > 1, a_i > 0, (b_i - x_i) > 0. \qquad (6.2.4)$$

$$U(X) = -\Sigma a_k^{\left(\frac{b_k - x_k}{a_k}\right)} \qquad a_i > 0. \qquad (6.2.5)$$

$$U(X) = \min\left\{\frac{x_k - b_k}{a_k}\right\} \qquad a_i > 0. \qquad (6.2.6)$$

The utility functions considered in Section 6.1 are special cases of (6.2.1), (6.2.2), (6.2.3) and (6.2.6), which correspond to these functions when all of the b's are 0.

The demand functions corresponding to (6.2.1) are of the form

$$h^i(P,\mu) = b_i - \frac{a_i}{p_i} \Sigma b_k p_k + \frac{a_i}{p_i} \mu \,. \tag{6.2.7}$$

This is the well-known Klein-Rubin [1947] linear expenditure system. The utility function (6.2.1) is a translated Cobb-Douglas. Similarly, (6.2.2), (6.2.3) and (6.2.6) are translations of the C.E.S. and fixed-coefficient cases considered in Section 6.1. The demand functions corresponding to (6.2.2), (6.2.3) and (6.2.4) are given by

$$h^i(P,\mu) = b_i - \gamma_i(P) \Sigma b_k p_k + \gamma_i(P)\mu \tag{6.2.8}$$

where

$$\gamma_i(P) = \frac{\left(\dfrac{p_i}{a_i}\right)^{\frac{1}{c-1}}}{\Sigma p_k \left(\dfrac{p_k}{a}\right)^{\frac{1}{c-1}}} \,.$$

The utility function corresponding to $c > 1$, (6.2.4), is not a generalization of an admissible C.E.S. case, but it includes the familiar additive quadratic ($c = 2$). The demand functions corresponding to (6.2.5) are given by

$$h^i(P,\mu) = b_i - \frac{a_i \Sigma p_k b_k}{\Sigma p_k a_k} + \frac{a_i \mu}{\Sigma p_k a_k} - a_i \log p_i + \frac{a_i \Sigma p_k a_k \log p_k}{\Sigma p_k a_k} \,.$$

The income-consumption curves are parallel straight lines. The demand functions corresponding to the translated fixed coefficient case, (6.2.6), are given by

$$h^i(P,\mu) = b_i - [\frac{a_i}{\Sigma a_k p_k}] b_k p_k + [\frac{a_i}{\Sigma a_k p_k}]\mu \,. \tag{6.2.10}$$

Since the demand functions are locally linear in expenditure, the indirect utility functions are of the form

$$\Psi(P,\mu) = \frac{\mu}{g(P)} - \frac{f(P)}{g(P)} \qquad (6.2.11)$$

where $f(P)$ and $g(P)$ are homogeneous of degree 1. For the linear expenditure system, (6.2.1),

$$g(P) = \pi \, p_k^{a_k} \qquad (6.2.12)$$

and

$$f(P) = \Sigma \, b_k p_k \, . \qquad (6.2.13)$$

In the three C.E.S.-like cases, (6.2.2), (6.2.3) and (6.2.4),

$$g(P) = [\Sigma \, a_k^{\frac{-1}{c-1}} p_k^{\frac{1}{c-1}}]^{\frac{c-1}{c}} \qquad (6.2.14)$$

and $f(P)$ is given by (6.2.13). In the case of parallel income-consumption curves, (6.2.5),

$$g(P) = \Sigma \, a_k p_k \qquad (6.2.15)$$

$$f(P) = (\Sigma \, a_k p_k)(\log \Sigma \, a_k p_k) + \Sigma \, b_k p_k - \Sigma \, a_k p_k \log p_k \, . \qquad (6.2.16)$$

Finally, in the translated fixed-coefficient case, (6.2.6), $f(P)$ is given by (6.2.13) and $g(P)$ by (6.2.15).

Solving the indirect utility function, (6.2.11), for μ, we find that the expenditure function is of the form

$$\mu = E(P,s) = f(P) + g(P)s \, . \qquad (6.2.17)$$

Hence, the cost of living index is of the form

$$I(P^a, P^b, s, R) = \frac{f(P^a) + g(P^a)s}{f(P^b) + g(P^b)s}.$$ (6.2.18)

It is often more useful to specify the base indifference curve in terms of base income m and to write the cost of living index in the form

$$I^*(P^a, P^b, m, R) = \alpha(P^a, P^b) + \frac{1}{m}\, \beta(P^a, P^b)$$ (6.2.19)

where

$$\alpha(P^a, P^b) = \frac{g(P^a)}{g(P^b)}$$ (6.2.20a)

$$\beta(P^a, P^b) = f(P^a) - f(P^b)\frac{g(P^a)}{g(P^b)}.$$ (6.2.20b)

Two cases are of special interest.

Theorem: The cost of living index corresponding to the linear expenditure system, (6.2.1), is of the form

$$I(P^a, P^b, s, R) = \frac{\Sigma\, b_k p_k^a + s\pi(p_k^a)^{a_k}}{\Sigma\, b_k p_k^b + s\pi(p_k^b)^{a_k}}.$$ (6.2.21)

This result was the conclusion of the article by Klein and Rubin [1947] in which they introduced the linear expenditure system. Interestingly enough, they entitled that paper "A Constant-Utility Index of the Cost of Living."

Theorem: If the direct utility function is of the translated fixed-coefficient form (6.2.6), then the cost of living index $I(P^a, P^b, s^b, R)$ is equal to its Laspeyres upper bound $J(P^a, P^b, X^b)$; the cost of living index $I(P^a, P^b, s^a, R)$ is equal to its Paasche lower bound $J(P^a, P^b, X^a)$.

6.3 Quadratic direct utility function

The quadratic utility function is best treated in matrix form. Let B denote an n x 1 vector and A and n x n matrix. The direct quadratic utility function is given by

$$U(X) = X'B - \frac{1}{2}X'AX \tag{6.3.1}$$

where X is an n x 1 vector of commodities and primes denote transpose. We do not explicitly specify the regularity conditions for this utility function but restrict ourselves to a region of the commodity space in which it is well behaved. We also ignore the problems posed by goods which are not consumed; we work in a region of the price-expenditure space in which the set of goods consumed remains unchanged. As Wegge [1968] shows, this is not a trivial restriction in the case of the direct quadratic.

It is easy to verify that the demand functions corresponding to (6.3.1) are of the form

$$X = A^{-1}B + A^{-1}P \left[\frac{\mu - P'A^{-1}B}{P'A^{-1}P} \right], \tag{6.3.2}$$

where the expression in brackets is a scalar. Thus, the implied demand functions are locally linear in expenditure. The indirect utility function is given by

$$-\frac{1}{2} \frac{(\mu - P'A^{-1}B)^2}{P'A^{-1}P}.$$

If $P'A^{-1}P$ is positive within the admissible region of the price-expenditure space, we rewrite the indirect utility function in its "Gorman form":

$$\Psi(P,\mu) = \frac{\mu - P'A^{-1}B}{\sqrt{P'A^{-1}P}} \tag{6.3.3a}$$

If $P'A^{-1}P$ is negative, then

$$\Psi(P,\mu) = \frac{\mu - P'A^{-1}B}{\sqrt{-P'A^{-1}P}}. \tag{6.3.3b}$$

The corresponding expenditure function is given by

$$\mu = P'A^{-1}B + s\sqrt{\pm\, P'A^{-1}P}$$

and the cost of living index by

$$I(P^a,P^b,s,R) = \frac{P^{a'}A^{-1}B + s\sqrt{\pm\, P^{a'}A^{-1}P^a}}{P^{b'}A^{-1}B + s\sqrt{\pm\, P^{b'}A^{-1}P^b}}. \tag{6.3.4}$$

If $B = 0$, the direct utility function is a "homogeneous quadratic," and the cost-of-living index is independent of the base indifference curve:

$$I(P^a,P^b,s,R) = \sqrt{\frac{P^{a'}A^{-1}P^a}{P^{b'}A^{-1}P^b}}. \tag{6.3.5}$$

The homogeneous quadratic is of particular importance because it corresponds to Irving Fisher's Ideal Index, the geometric mean of a Laspeyres and a Paasche:

$$[J(P^a,P^b,H(P^b,\mu^b;R))J(P^a,P^b,H(P^a,\mu^a;R))]^{\frac{1}{2}}.$$

Furthermore, the homogeneous quadratic is the only preference ordering for which this is true:

Theorem: The cost of living index coincides with Fisher's Ideal Index

$$I^*(P^a,P^b,m,R) = [J(P^a,P^b,H(P^b,\mu^b;R))J(P^a,P^b,H(P^a,\mu^a;R))]^{\frac{1}{2}}$$

if and only if the preference ordering is a homogeneous quadratic.[11]

Proof: If the direct utility function is a homogeneous quadratic, then the Laspeyres and Paasche indexes are given by

$$J[P^a,P^b,H(P^b,\mu^b;R)] = \frac{P^{a'}A^{-1}P^b}{P^{b'}A^{-1}P^b}$$

$$J[P^a,P^b,H(P^a,\mu^a;R)] = \frac{P^{a'}A^{-1}P^a}{P^{b'}A^{-1}P^a}$$

Since $P^{a'}A^{-1}P^b = P^{b'}A^{-1}P^a$, the ideal index is (6.3.5).

If the cost of living index coincides with the ideal index, then the Laspeyres index is independent of μ^b, and hence the demand functions exhibit expenditure proportionality. We write the indirect function in its Gorman form, $\Psi(P,\mu) = \mu/\phi(P)$, where $\phi(P)$ is a function homogeneous of degree 1. The demand functions are given by $h^i(P,\mu) = \phi_i(P)\mu/\phi(P)$. The cost of living index I is implicitly defined by

$$\Psi(P^b,m) = \Psi(P^a,mI)$$

so

$$\frac{\phi(P^a)}{\phi(P^b)} = I .$$

The product of the Laspeyres and Paasche indexes can be written as

$$\frac{\Sigma\, h^k(P^b,\mu^b)p_k^a}{\Sigma\, h^k(P^b,\mu^b)p_k^b}\ \frac{\Sigma\, h^k(P^a,\mu^a)p_k^a}{\Sigma\, h^k(P^a,\mu^a)p_k^b} = \frac{\Sigma\, \phi_k(P^b)p_k^a}{\Sigma\, \phi_k(P^b)p_k^b}\ \frac{\Sigma\, \phi_k(P^a)p_k^a}{\Sigma\, \phi_k(P^a)p_k^b}$$

$$= \frac{\Sigma\, \phi_k(P^b)p_k^a}{\phi(P^b)}\ \frac{\phi(P^a)}{\Sigma\, \phi_k(P^a)p_k^b}.$$

If the cost of living index is equal to the ideal index, then

$$\left(\frac{\phi(P^a)}{\phi(P^b)}\right)^2 = \frac{\Sigma\, \phi_k(P^b)p_k^a}{\phi(P^b)}\ \frac{\phi(P^a)}{\Sigma\, \phi_k(P^a)p_k^b}$$

or, equivalently,

$$\phi(P^a)\ \Sigma\, \phi_k(P^a)p_k^b = \phi(P^b)\ \Sigma\, \phi_k(P^b)p_k^a .$$

Differentiating with respect to p_i^a yields

$$\phi_i(P^a)\ \Sigma\, \phi_k(P^a)p_k^b + \phi(P^a)\ \Sigma\, \phi_{ki}(P^a)p_k^b = \phi(P^b)\phi_i(P^b) .$$

Differentiating this with respect to p_j^b yields

$$\phi_i(P^a)\phi_j(P^a) + \phi(P^a)\phi_{ji}(P^a) = \phi_j(P^b)\phi_i(P^b) + \phi(P^b)\phi_{ij}(P^b) .$$

Since the right hand side and the left hand side are equal regardless of the values of P^a and P^b, both must be independent of P and hence constant:

$$\phi_i(P)\phi_j(P) + \phi(P)\phi_{ji}(P) = c_{ij} .$$

Multiplying by p_j and summing over j yields

$$\phi_i(P)\phi(P) = \sum_j p_j c_{ij}$$

since the homogeneity of ϕ implies

$$\sum_j p_j \phi_{ji}(P) = 0 .$$

Multiplying by p_i and summing over i yields

$$[\phi(P)]^2 = \sum_i \sum_j p_i p_j c_{ij}$$

so

$$\phi(P) = \pm \sum_i \sum_j p_i p_j c_{ij}$$

which is the homogeneous quadratic.

We remark that if $c_{ij} = a_i a_j$, then

$$\phi(P) = \sum a_k p_k$$

and that

$$\Psi(P,\mu) = \frac{\mu}{\sum a_k p_k}$$

is the indirect utility function corresponding to the homogeneous fixed-coefficient utility function. It should be no surprise that the homogeneous fixed-coefficient case appears here, for this is the case in which the cost of living index coincides with both the Laspeyres and the Paasche indexes, and, hence, it must coincide with their geometric mean.

7. A Preference Field Quantity Index

As we have seen, the cost of living index provides a precise answer to a narrow and specific question. If one wishes to compare expenditures required to attain a particular base indifference curve at two sets of prices, then, by definition, the cost of living index is the appropriate index. But price indexes are often used to deflate an index of total expenditure to obtain an index of quantity or "real consumption." With less logic but the same purpose they are used to deflate indexes of money income to obtain indexes of real income or money wages to obtain real wages. Although the last two cases are clouded by problems involving saving and the labor-leisure choice, the purposes of these indexes is the measure "quantity."[12] In this section we show how a quantity index can be constructed by a procedure analogous to that used to construct the cost of living index. We call such an index a "preference field quantity index" to distinguish it from other types of quantity indexes and to suggest its relation to the "preference field price index" or cost of living index of Section 2. Before constructing the preference field quantity index, we give a careful summary of the logic which lies behind the preference field price index. In Section 8 we examine the conditions under which the preference field quantity index coincides with the quantity index obtained by using the cost of living index to deflate an index of expenditure.

7.1 The preference field price index

Given a preference ordering R and a base indifference curve s we defined the cost of living index by

$$I(P^a, P^b, s, R) = \frac{c_a}{c_b} \tag{7.1.1}$$

where c_a and c_b are implicitly defined by

$$s = \Psi(P^a, c_a; R)$$

$$s = \Psi(P^b, c_b; R) .$$

In theoretical work involving indirect utility functions, it is standard practice to work with "normalized prices." We define y_i, the normalized price of the ith good, by $y_i = p_i/\mu$ and let Y denote the corresponding vector. The ordinary demand functions $h^i(P,\mu;R)$ can be written as $g^i(Y;R)$ since they are homogeneous of degree 0 in all prices and expenditure:

$$g^i(Y;R) = h^i(Y,1;R) = h^i (\frac{p_1}{\mu},..., \frac{p_n}{\mu},1;R) = h^i(P,\mu;R) .$$

Similarly, the indirect utility function $\Psi(P,\mu;R)$ can be written as $\phi(Y;R)$ since it too is homogeneous of degree 0 in all prices and expenditure:

$$\phi(Y;R) = \Psi(Y,1;R) = \Psi (\frac{p_1}{\mu},..., \frac{p_n}{\mu},1;R) = \Psi(P,\mu;R) .^{13}$$

The ordinary demand functions are related to the normalized indirect utility function by

$$g^i(Y) = h^i(Y,1) = \frac{\phi_i(Y)}{\Sigma \, y_k \phi_k(Y)}.$$

Using this new notation, the cost of living index is given by (7.1.1) where c_a and c_b are implicitly defined by

$$s = \phi (\frac{1}{c_a}P^a;R)$$

$$s = \phi (\frac{1}{c_b}P^b;R) .$$

We can give a straightforward interpretation of the cost of living index in terms of the indifference curves corresponding to the indirect utility function $\phi(Y)$. We remind the reader that in Figure 1 utility *increases* as you move *toward* the origin. We begin with the base preferences ordering (represented by the indifference map) and, from the map, the base indifference curve, which we denote by s. We let $Y^a = P^a$ and $Y^b = P^b$ denote the comparison and reference prices, respectively. In general, neither the reference nor the comparison prices will lie on the base indifference curve. We let $Y^{a'}$ denote the point at which the ray from the origin through Y^a intersects the base indifference curve. Similarly, $Y^{b'}$ denotes the intersection of the base indifference curve with the ray from the origin through Y^b

It is instructive to decompose the construction of the cost of living index comparing Y^a with Y^b using the base indifference curve s into three separate parts: a comparison of Y^a with $Y^{a'}$, with $Y^{b'}$, and a comparison of $Y^{b'}$ with Y^b. First, we compare Y^a with $Y^{a'}$. That is, suppose the comparison prices Y^a and the reference prices (for our present purpose, $Y^{a'}$) lie on the same ray: $Y^a = c_a Y^{a'}$. Then the cost of living index $I(Y^a, Y^{a'}, s, R)$ is equal to c_a, a result which is intuitively obvious and natural. Formal justification is provided by

$$s = \phi(\frac{1}{c_a} Y^a; R) = \phi(Y^{a'}; R).$$

If, for example, $Y^a = 4Y^{a'}$, and $Y^{a'}$ is on the base indifference curve, then $I(Y^a, Y^{a'}, s, R) = 4$.

We now compare $Y^{a'}$ with $Y^{b'}$. That is, suppose the comparison prices are $Y^{a'}$ and the reference prices $Y^{b'}$. Since, by construction, both $Y^{a'}$ and $Y^{b'}$ lie on the base indifference curve, the value of index is 1: $I(Y^{a'}, Y^{b'}, s, R) = 1$.

Finally, we compare $Y^{b'}$ with Y^b. In this case the comparison prices $Y^{b'}$ and the reference prices Y^b lie on the same ray: $Y^b = c_b Y^{b'}$. Hence, the cost-of-living index $I(Y^{b'}, Y^b, s, R)$ is equal to $1/c_b$, as common sense requires. For example, if $Y^b = 3Y^{b'}$, then $I(Y^{b'}, Y^b, s, R) = \frac{1}{3}$

We must now combine these three results to obtain the cost-of-living index we originally sought: $I(Y^a, Y^b, s, R)$. It is at this stage that our argument loses some of its intuitive appeal. It is perfectly clear how we ought to compare two-price situations which lie on a common ray or two-price situations which lie on the base indifference curve, but it is less clear how we ought to proceed in other cases. Formally, we can proceed by introducing a formal rule which enables us to combine indexes with a common base indifference curve:

$$I(Y^a,Y^c,s,R) = I(Y^a,Y^b,s,R)I(Y^b,Y^c,s,R) .$$

We saw in Section 2 that cost-of-living indexes can be combined in this way. But the intuitive justification for this step is not a completely comfortable one. To understand the way in which this multiplication rule operates, suppose that the reference prices lie on the base indifference curve but not on the same ray as the comparison prices. The index is then equal to the ratio of the comparison prices to the point where the comparison ray intersects the base indifference curve. In effect, we construct the index as if the reference prices were at this intersection. Our rationale for this is that the reference prices and the intersection lie on the base indifference curve, and we regard points on the base indifference curve as "equivalent".

We now take account of the fact that the reference prices Y^b need not lie on the base indifference curve. We do this by making use of $P^{b'}$, the point at which the reference ray intersects the base indifference curve. As we saw, if $Y^b = c_b Y^{b'}$, then $I(Y^{b'},Y^b,s,R) = 1/c_b$. Furthermore, if the comparison prices lie anywhere on the base indifference curve, the value of the index would still be $1/c_b$. If neither the comparison nor the reference prices lie on the base indifference curve, we construct the index by reducing both reference and comparison prices to the base indifference curve and making use of the "equivalence" of all price situations on the base indifference curve. Thus, in our example, the value of the index would be

$$I(Y^a,Y^b,s,R) = I(Y^a,Y^{a'},s,R)I(Y^{a'},Y^{b'},s,R)I(Y^{b'},Y^b,s,R) = 4 \times 1 \times \frac{1}{3} = \frac{4}{3} .$$

The procedure for constructing a cost-of-living index based on a particular indifference curve s can be summarized in three simple axioms:

$$\text{If} \quad s = \phi(Y^a;R) \quad \text{and} \quad s = \phi(Y^b;R), \quad \text{then } I(Y^a,Y^b,s,R) = 1 . \qquad \text{A1.}$$

Figure 1

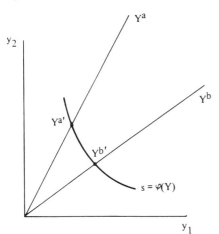

Figure 2

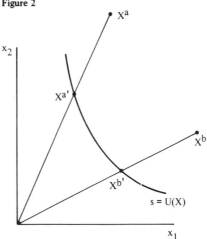

That is, if the comparison and the reference prices both lie on the base indifference curve, then the value of the index is 1.

$$\text{If} \quad s = \phi(Y^b;R) \quad \text{and} \quad Y^a = \lambda Y^b, \quad \text{then} \quad I(Y^a,Y^b,s,R) = \lambda . \qquad \text{A2.}$$

In words, if the reference prices lie on the base indifference curve and the comparison prices lie on the same ray as the reference prices, then the value of the index is the factor of proportionality relating the comparison to the reference prices.

$$I(Y^a,Y^b,s,R)I(Y^b,Y^c,s,R) = I(Y^a,Y^c,s,R) . \qquad \text{A3.}$$

If two cost-of-living indexes are based on the same base indifference curve, and the reference prices in the first are the same as the comparison prices in the second, then the product of these two indexes is the cost-of-living index whose comparison prices are those of the first and whose reference prices are those of the second. Irving Fisher [1922, p.270] called this the "circular" property.

This **axiomatic** treatment of the procedure for constructing the cost-of-living index makes it clear that the base indifference map plays no role and that the base curve does all the work. It is sometimes useful to discard the original indifference map and work only with the base indifference curve. Sometimes, however, it is useful to go a step further and think of the index as being constructed from a pseudo indifference map which is defined as the radial or homothetic blowup of the base indifference curve. Unless the original indifference map was homothetic to the origin, the pseudo map does not coincide with the original map. One advantage of introducing the pseudo map is that because it is homothetic to the origin, the cost-of-living index is independent of which pseudo indifference curve is treated as the base. In particular, without loss of generality, we may use the curve on which the reference prices lie as the base curve. On the pseudo map (but not on the original map unless it was homothetic) if the reference prices and the comparison prices lie on the same curve, then the value of the index is 1. Notice, however, that if the reference and comparison prices lie on the same indifference curve on the original map, and their common curve is not the base curve, then the value of the index need not be 1.

7.2 Preference field quantity indexes

We now use the same formal procedure to define a preference field quantity index as we used to define a preference field price index or cost of living index. That is, we define the preference field quantity index $Q(X^a, X^b, s, R)$ by

$$Q(X^a, X^b, s, R) = \frac{\theta_a}{\theta_b}$$

where θ_a and θ_b are defined by

$$s = U\left(\frac{1}{\theta_a} X^a\right)$$

and

$$s = U\left(\frac{1}{\theta_b} X^b\right).$$

The verbal and graphical interpretation of the preference field quantity index is essentially the same as that of the preference field price index. We begin by choosing a base preference ordering and from it a base indifference curve. In general, neither the comparison quantities X^a nor the reference quantities X^b lie on the base indifference curve. Instead of comparing X^a directly with X^b, we compare X^a with a collection of goods $X^{a\,'}$ which lies on s and which is proportional to X^a. Similarly, we compare X^b with a collection $X^{b\,'}$ which lies on s and which is proportional to X^b. Graphically, the preference field quantity index can be represented by a diagram, Figure 2, identical to that used to illustrate the construction of the cost of living index.

Since the preference field quantity index is less familiar than the cost of living index, it is useful to examine the meaning of this index. Formally, the axioms satified by this index are identical with those satisfied by the preference field price index:

$$\text{If } s = U(X^a) \text{ and } s = U(X^b) \text{ then } Q(X^a,X^b,s,R) = 1 \qquad \text{A1.}$$

$$\text{If } s = U(X^b) \text{ and } X^a = \lambda X^b \text{ then } Q(X^a,X^b,s,R) = \lambda \qquad \text{A2.}$$

$$Q(X^a,X^b,s,R)Q(X^b,X^c,s,R) = Q(X^a,X^c,s,R) . \qquad \text{A3.}$$

That is, (A1.) if both the comparison and reference quantities lie on the base indifference curve, then the value of the index is 1. (A2.) If the reference quantities lie on the base indifference curve, and the comparison quantities and the reference quantities lie on a common ray, then the value of the index is the ratio of the comparison to the reference quantities. (A3.) If two indexes have the same base, and the reference quantities of the first are equal to the comparison quantities of the second, then the product of the two indexes is an index whose comparison quantities are equal to those of the first and whose reference quantities are those of the second. These three axioms uniquely determine the preference field quantity index corresponding to any base indifference curve.

As with the preference field price index, the construction of the index depends only on the base indifference curve but not on the underlying preference ordering or indifference map. It is useful to consider the index in relation to a pseudo indifference map constructed as the homothetic blowup of the base indifference curve. On the pseudo indifference map the index is independent of the choice of the pseudo indifference curve used as the base. Thus, we can choose the curve corresponding to the reference quantities with no loss of generality. If the comparison and reference quantities lie on the same pseudo indifference curve, then the value of the index is 1. But if the comparison and reference quantities lie on the same indifference curve on the original map, the value of the index need not be unity unless they both lie on the base indifference curve. This is important because it underscores the fact that the preference field quantity index does not compare the levels of utility corresponding to the reference and comparison quantities.

8. Price, Quantity, and Expenditure Indexes

Since expenditure is a scalar, an expenditure index $M(\mu_a, \mu_b)$ is naturally defined by

$$M(\mu_a, \mu_b) = \frac{\mu_a}{\mu_b}.$$ Such an index is independent of preferences. In this section we examine

the relationships among price, quantity, and expenditure indexes. In particular, we are interested in the conditions under which the expenditure index $M(\mu^a, \mu^b)$ can be decomposed into the product of a price index and a quantity index or, equivalently, the conditions under which the quantity index is equal to the expenditure index deflated by a price index or, equivalently, the conditions under which the price index is equal to the expenditure index deflated by a quantity index. The answers to these questions depend on the types of price and quantity indexes which are admissible.

One approach is to require the price index to be a cost of living index and to define the quantity index by

$$Q(P^a, P^b, \mu^a, \mu^b, s, R) = \frac{M(\mu^a, \mu^b)}{I(P^a, P^b, s, R)}.$$

Alternatively, if we insist that the quantity index be a preference field index, we can define the price index by

$$I(X^a, X^b, \mu^a, \mu^b, s, R) = \frac{M(\mu^a, \mu^b)}{Q(X^a, X^b, s, R)}.$$

The drawback of this approach is that the derived indexes do not satisfy the axioms discussed in Section 7 unless the base preference ordering is homothetic. Furthermore, it is easy to verify that

$$Q(P^a,P^b,\mu^a,\mu^b,s,R)Q(P^b,P^c,\mu^b,\mu^c,s,R) = Q(P^a,P^c,\mu^a,\mu^c,s,R)$$

so that A3., the least intuitively appealing of our axioms, is always satisfied by a quantity index defined in this way. Hence, failure of the quantity index to satisfy our axioms implies a violation of either A1., which requires that if both the comparison and reference quantities lie on the base indifference curve, then the value of the index is 1, or A2., which requires that if the reference quantities lie on the base indifference curve and both the comparison and reference quantities lie on a common ray, then the index is the ratio of the comparison to the reference quantities. Both of these axioms have such strong intuitive appeal that a quantity index which violates them has little conceptual appeal. We show in this section that the preference field quantity index is equal to the expenditure index deflated by the cost of living index if and only if the preference ordering is homothetic to the origin. This implies that unless the preference ordering is homothetic, $Q(P^a,P^b,\mu^a,\mu^b,S,R)$ violates either A1. or A2.

We now require both the price and the quantity indexes to be preference field indexes and determine the class of preference orderings for which their product is equal to the expenditure index.

Theorem: $M(\mu^a,\mu^b) = I(P^a,P^b,s,R)Q[H(P^a,\mu^a;R),H(P^b,\mu^b;R),s,R]$, for all P^a,P^b,μ^a,μ^b, if and only if the preference ordering is homothetic to the origin.

The statement of the theorem requires that the result hold for only one base indifference curve, s, but the conclusion implies that if it holds for one base curve, then the preference ordering is such that it holds for every base curve.

Proof: If the preference ordering is homothetic to the origin, and the demand functions were generated by this preference ordering, we can write the direct utility function U(X) as a function homogeneous of degree 1; the indirect utility function $\Psi(P,\mu) = U[H(P,\mu)]$

$= U[H(P,1)]\mu$ can be written as $\dfrac{\mu}{\phi(P)}$ where $\phi(P)$ is homogeneous of degree 1, $\phi(P) = 1/U[H(P,1)]$.

The cost of living index is given by

$$I(P^a,P^b,s,R) = \frac{\phi(P^a)}{\phi(P^b)}.$$

The preference field quantity index is given by

$$Q(X^a,X^b,s,R) = \frac{\theta_a}{\theta_b} = \frac{U(X^a)}{U(X^b)}$$

since

$$s = U\left(\frac{1}{\theta_a}X^a\right) = \frac{1}{\theta_a}U(X^a)$$

and

$$s = U\left(\frac{1}{\theta_b}X^b\right) = \frac{1}{\theta_b}U(X^b).$$

But $U(X^a) = \dfrac{\mu^a}{\phi(P^a)}$ and $U(X^b) = \dfrac{\mu^b}{\phi(P^b)}$, so the product of the preference field price and quantity indexes is μ^a/μ^b.

We prove the second part of the theorem by showing that if the expenditure index is equal to the product of the preference field price and quantity indexes, then the demand functions exhibit expenditure proportionality. We first observe that the quantity index must be homogeneous of degree 1 in μ^a since μ^a appears only in $M(\mu^a,\mu^b)$ and in $H(P^a,\mu^a)$. Let P^a be any price vector; set P^b equal to P^a. Choose μ^b so that $\Psi(P^b,\mu^b) = s$. Then $X^b = H(P^b,\mu^b)$ lies on the base indifference curve. If we set μ^a equal to μ^b, then $X^a =$

X^b and $Q(X^a, X^b, s, R) = 1$. Suppose $\mu^a = \lambda \mu^b$. Since the quantity index is homogeneous of degree 1 in λ, its value is λ. This means that X^a lies on the pseudo indifference curve which is a radial blowup of the base indifference curve by the scale factor λ. But X^a must lie in the feasible set defined by P^a, μ^a. This feasible set is a radial blowup of the feasible set defined by P^b, μ^b. If the indifference curves are strictly convex, X^b is the only point on the base indifference curve in the feasible set of P^b, μ^b. Hence, λX^b is the only point on the radial blowup of the base indifference curve which lies in the feasible set of P^a, μ^a. Hence, $X^a = \lambda X^b$ and the demand functions exhibit expenditure proportionality.

It might be argued that this theorem is not surprising since we have not allowed the base indifference curve to vary with μ^a or μ^b. If the base indifference curve were always chosen so that $s = \Psi(P^b, \mu^b)$, perhaps some non-homothetic preference ordering would suffice. We now show that this is not the case.

Theorem: $M(\mu^a, \mu^b) = I[P^a, P^b, \Psi(P^b, \mu^b; R); R] Q[H(P^a, \mu^a; R), H(P^b, \mu^b; R), \Psi(P^b, \mu^b; R), R]$, for all P^a, P^b, μ^a, μ^b if and only if the preference ordering is homothetic to the origin.

That is, allowing the base of the index to vary with μ^b does not permit any generalization of our theorem; clearly, allowing it to vary with μ^a would not.

Proof: The proof of the previous theorem showed that if the indifference map is homothetic, both the price and quantity indexes are independent of the base indifference curve, and their product is equal to the expenditure index.

To prove that only homothetic preference orderings will work, we first observe that the quantity index is homogeneous of degree 1 in μ^a. We proceed as in the proof of the previous theorem except that we are free to choose any initial value for μ^b.

It might still be objected (by analogy with the fact that the product of a Laspeyres price index and a Paasche quantity index is equal to the expenditure index) that we should not require the same base indifference curve to serve for both indexes. We now show that basing the price index on the reference situation and the quantity index on the comparison situation does not permit any generalization of our result.

Theorem:

$M(\mu^a,\mu^b) = I[P^a,P^b,\Psi(P^b,\mu^b;R);R]Q[H(P^a,\mu^a;R),H(P^b,\mu^b;R),\Psi(P^a,\mu^a;R),R]$ for all P^a,P^b,μ^a,μ^b if and only if the preference ordering is homothetic to the origin.

Proof: We have already seen that if the indifference map is homothetic, the required result holds.

To prove that only a homothetic preference ordering will work, we show that the demand functions exhibit expenditure proportionality. Instead of focusing on the comparison quantities, $H(P^a,\mu^a)$, we focus on the reference quantities, $H(P^b,\mu^b)$. This is simpler because μ^a appears twice in the quantity index while μ^b appears only once. Let P^b be any price vector, and set P^a equal to P^b. Choose μ^a arbitrarily to fix the base indifference curve of the quantity index. Since $P^a = P^b$, the cost-of-living index is 1, and hence the quantity index is equal to μ^a/μ^b for all μ^a,μ^b. Hence, it is homogeneous of degree -1 in μ^b. A slight modification of the argument used in our first proof establishes that $X^b = \lambda X^a$, so the demand functions exhibit expenditure proportionality.

9. Price Indexes for Demand Analysis

Demand theory tells us that the demand for a good is a function of its own price, the prices of all other goods, and total expenditure. Without additional assumptions, the theory says very little about the form of the demand functions. To estimate demand functions, specific assumptions must be made about their functional form. It is sometimes assumed that the demand for each good is a function of its own price and expenditure where these variables have been deflated by an appropriate price index. That is

$$h^i(P,\mu) = g^i \left[\frac{p_i}{T(P)}, \frac{\mu}{T(P)} \right] \qquad (9.1)$$

where the price index $T(P)$ is assumed to be homogeneous of degree 1. The same index is used to deflate both price and expenditure and appears in every demand equation. This means that p_i appears twice in the demand function for the ith good: in its own right as a price variable and again as an argument of the index function T. If the demand functions are of this form, the prices of "other goods" enter the demand functions only through the index function.

In Pollak [1972] I defined "generalized additive separability" as follows: a system of demand functions exhibits generalized additive separability if its demand functions are of the form

$$h^i(Y) = \Gamma^i[y_i, R(Y)], \qquad i = 1,\ldots,n .$$

That is, the demand for each good is a function of its own normalized price and an index function which depends on all normalized prices. The same index function appears in the demand function for every good.

Theorem: A system of demand functions is of the form (9.1) if and only if it exhibits generalized additive separability and the index function $R(Y)$ is homothetic.

Proof: If the demand functions are (9.1), we can define $\bar{g}^i(.,.)$ by

$$\bar{g}^i\left[\frac{p_i}{\mu}, \frac{T(P)}{\mu}\right] = g^i\left[\frac{p_i}{T(P)}, \frac{\mu}{T(P)}\right]$$

since the original arguments of g^i can be recovered from those of $\bar{g}^i$. Since the demand functions are homogeneous of degree 0 in P and μ,

$$\bar{g}^i\left[\frac{p_i}{\mu}, \frac{T(P)}{\mu}\right] = \bar{g}^i\left[\frac{\lambda p_i}{\lambda\mu}, \frac{T(\lambda P)}{\lambda\mu}\right] = \bar{g}^i\left[\frac{p_i}{\mu}, \frac{T(\lambda P)/\lambda}{\mu}\right] = \bar{g}^i[y_i, T(Y)] ,$$

so the demand functions exhibit generalized additive separability where $T(Y)$ is homogeneous of degree 1.

If $h^i(Y) = \Gamma^i[y_i, R(Y)]$, where R is homothetic, we can redefine R and Γ so R is homogeneous of degree 1. Then

$$h^i(P,\mu) = \Gamma^i \left[\frac{p_i}{\mu} , \frac{R(P)}{\mu}\right] = g^i_i \left[\frac{p_i}{R(P)} , \frac{\mu}{R(P)}\right]$$

since the original arguments of Γ^i can be recovered from those of g^i.

Fourgeaud and Nataf [1959] explicitly characterize the systems of demand functions of the form (9.1) which satisfy the budget constraint and the Slutsky symmetry conditions and hence are theoretically plausible in the sense that they can be derived from a well-behaved preference ordering. Their results are summarized in Pollak [1972, Section IId]. There are four principal cases:

$$h^i(Y) = \frac{\Psi^i[\log y_i - \log R(Y)]}{y_i} \tag{9.2}$$

where $\Psi^i(.)$ is a function of a single variable, and R is defined implicitly by

$$\Sigma \, \Psi^k[\log y_k - \log R] = 1.$$

$$h^i(Y) = \frac{a_i R(Y) + b_i[1 - R(Y)]}{y_i} \tag{9.3}$$

where R(Y) is homogeneous of degree 1 in Y and

$$R(Y) = T[\Sigma \, a_k \log y_k, \Sigma \, b_k \log y_k], \qquad \Sigma \, a_k = \Sigma \, b_k = 1. \,^{14}$$

The other two cases are

$$h^i(Y) = \frac{\delta[T(Y)][\alpha_i \log y_i - \alpha_i \log T(Y) + \beta_i] + \alpha_i}{y_i}, \tag{9.4}$$

$$\Sigma \, \alpha_k = 1, \Sigma \, \beta_k = 0$$

where $\delta(.)$ is a function of a single variable and

$$T(Y) = \pi \, y_k^{\alpha_k}$$

and

$$h^i(Y) = \frac{\delta[T(Y)][\beta_i - \alpha_i y_i^c \, T(Y)^{-c}] + \beta_i}{y_i}, \quad \Sigma \, \beta_k = 1 \qquad (9.5)$$

where $\delta(.)$ is a function of a single variable and

$$T(Y) = [\Sigma \, \alpha_k y_k^c]^{\frac{1}{c}} .$$

The first case, (9.2), exhibits expenditure proportionality and hence is not very interesting for empirical demand analysis. In (9.3) the demand functions are locally linear in expenditure; the index function is a function of two Cobb-Douglas functions. In (9.4) and (9.5) the index functions are Cobb-Douglas and C.E.S., respectively.

Thus, the price indexes which appear in the demand functions depend in a specific way on the parameters of the preference ordering. There is no presumption that these price indexes coincide with the cost-of-living index; indeed, it is not entirely clear what the assertion that T(P) coincides with the cost-of-living index $I(P^a,P^b,s,R)$ would mean. To indicate that the prices in the demand functions are variables, we rewrite (9.1) as

$$h^i(P^t,\mu^t) = g^i \left[\frac{p_i^t}{T(P^t)} , \frac{\mu^t}{T(P^t)} \right] . \qquad (9.1')$$

We may interpret P^t as the price vector of period t although this is not essential. The cost-of-living index $I(P^t,P^b,s,R)$ is defined by

$$I(P^t,P^b,s,R) = \frac{E(P^t,s,R)}{E(P^b,s,R)} .$$

Two difficulties are immediately apparent: the first is that the cost-of-living index depends on the reference prices P^b as well as on P^t while the demand functions depend only on the comparison prices P^t. Since the reference prices (and also the base indifference curve) are held constant, the cost-of-living index is proportional to $E(P^t,s,R)$; the constant factor of proportionality is $1/E(P^b,s,R)$, and it can be absorbed into the parameters of the demand functions. Hence, we interpret the assertion that the cost-of-living index coincides with $T(P)$ to mean that

$$h^i(P^t,\mu^t) = g^i [\frac{p_i^t}{E(P^t,s)}, \frac{\mu^t}{E(P^t,s)}].$$

The fact that the cost-of-living index depends on the choice of the base indifference curve (unless the demand functions exhibit expenditure proportionality) is a more serious problem. In particular, in the absence of expenditure proportionality the assertion that the cost-of-living index coincides with $T(P)$ must be interpreted to mean that there exists a base indifference curve s^* such that $T(P)$ is proportional to $E(P,s^*,R)$ where the factor of proportionality can be absorbed into $T(P)$.

The linear expenditure system

$$h^i(P,\mu) = b_i - \frac{a_i}{p_i} \Sigma b_k p_k + \frac{a_i}{p_i} \mu$$

provides a good illustration. These demand functions can be written as

$$h^i(P,\mu) = b_i - \frac{a_i}{(\frac{p_i}{\Sigma b_k p_k})} + a_i \frac{(\frac{\mu}{\Sigma b_k p_k})}{(\frac{p_i}{\Sigma b_k p_k})} \tag{9.6}$$

which belongs to the Fourgeaud-Nataf class. In fact, the linear expenditure system is a special case of (9.5) where $\delta(T) = -T$ and $c = 1$. The cost-of-living index corresponding to the linear expenditure system is given by (6.2.21). For $s = 0$ this becomes

$$I(P^t, P^b, 0, R) = \frac{\Sigma b_k p_k^t}{\Sigma b_k p_k^b}$$

which is the appropriate price index for demand analysis in this case.

In general, however, the cost-of-living index does not coincide with T(P). For example, the indirect utility function

$$\Psi(P, \mu) = -\frac{1}{\mu} \pi p_k^{a_k} + \Sigma (a_k - b_k) \log p_k, \qquad \Sigma a_k = \Sigma b_k = 1$$

is a special case of (9.4). The demand functions are given by

$$h^i(P, \mu) = \frac{a_i}{p_i} - \frac{(a_i - b_i)\mu^2}{p_i \pi p_k^{a_k}} = \frac{a_i \left(\frac{\mu}{T}\right)}{p_i \left(\frac{1}{T}\right)} - \frac{(a_i - b_i)\left(\frac{\mu}{T}\right)^2}{\left(\frac{p_i}{T}\right)}$$

where $T(P) = \pi p_k^{a_k}$. The expenditure function corresponding to this indirect utility function is

$$\mu = \frac{-\pi p_k^{a_k}}{s - \Sigma (a_k - b_k) \log p_k}$$

and there is no value of s (independent of prices) for which T(P) is proportional to E(P,s,R).

This is not too surprising. On reflection, there was no reason to expect T(P) to coincide with the cost-of-living index. Different price indexes are needed for different purposes, and the cost-of-living index should not be expected to play a role in demand analysis. Furthermore, even in those cases in which T(P) and the cost-of-living index coincide, this relationship is of no use in empirical demand analysis. The trouble is that we do not start out knowing the cost-of-living index. The cost-of-living index cannot be calculated until the unknown parameters of the demand system have been estimated, and the fact that the unknown price index T(P) is proportional to the unknown expenditure function E(P,s*,R) does nothing to simplify the estimation problem.

Of course, if we know the cost-of-living index, the situation is very different. In that case, regardless of whether the demand functions involve price indexes, it is possible to calculate the demand functions directly from the cost-of-living index. The cost-of-living index $I^*(P^t, P^b, m, R)$ contains all the information about preferences, and it is fairly straightforward to retrieve this information and find the implied demand functions. To do this we write

$$E[P^t, \Psi(P^b, m)] = m I^*(P^t, P^b, m, R) .$$

Differentiating with respect to p_i^t yields

$$f^i[P^t, \Psi(P^b, m)] = \frac{\partial E[P^t, \Psi(P^b, m)]}{\partial p_i^t} = m \frac{\partial I^*(P^t, P^b, m, R)}{\partial p_i^t} .$$

The ordinary demand functions $h^i(P^t, \mu^t, R)$ can be calculated by finding the value of m, m^t, for which

$$\mu^t = E[P^t, \Psi(P^b, m^t)] = m^t I^*(P^t, P^b, m^t, R) .$$

Then

$$h^i(P^t, \mu^t) = f^i[P^t, \Psi(P^b, m^t)] .$$

If we begin knowing only the cost-of-living index $I(P^t, P^b, s, R)$, then we cannot retrieve the ordinary demand functions. However, even in terms of the fundamental question which the cost-of-living index is designed to answer, it is not enough to know only $I(P^t, P^b, s, R)$. The difficulty is that unless we also know the indirect utility function $\Psi(P, \mu)$, or have other equivalent information, we have no way to interpret the numerical value of s. We cannot associate it with a particular collection of goods and services or with a particular level of expenditure at a particular set of prices. Hence, if we know $I(P^t, P^b, s, R)$ and have enough information to interpret it meaningfully as a cost-of-living index, we can calculate the implied demand functions.

To summarize: it is useful to consider systems of demand functions which involve price indexes, but the price indexes which are relevant for demand analysis are unlikely to coincide with the cost-of-living index. Even if they coincide, knowing this is of no help in empirical demand analysis because the cost-of-living index cannot be computed until the unknown parameters of the system of demand equations have been estimated. The cost-of-living index formula $I^*(P^t, P^b, m, R)$ contains enough information about preferences to enable us to calculate the demand functions. So, if the formula for the cost-of-living index is known, then the demand functions can be calculated from it, regardless of whether the demand functions involve price indexes.

We now turn to more practical questions about the use of price indexes in empirical demand analysis. Assuming that the demand functions depend on price indexes, (9.1), when will the index $T(P)$ be a linear function of prices? We have already seen that the linear expenditure system is of this form; we now characterize the entire class.

Theorem: If the demand functions are of the Fourgeaud-Nataf form (9.1)

$$h^i(P, \mu) = g^i [\frac{P_i}{T(P)}, \frac{\mu}{T(P)}], \qquad T(\lambda P) = \lambda T(P)$$

and if the index function $T(P)$ is linear in prices

$$T(P) = \Sigma \, w_k p_k \, ,$$

then, except for degenerate cases in which less than three w's are non-zero, the demand functions fall into two classes:

$$h^i(P, \mu) = \frac{\beta_i \mu}{p_i} + \frac{(1 - \Sigma \beta_k) \alpha_i \mu}{\Sigma \alpha_k p_k} \qquad (9.7)$$

and

$$h^i(P,\mu) = \frac{\beta_i \mu}{p_i} + \delta \left[\frac{\Sigma \, \alpha_k p_k}{\mu} \right] \left[\frac{\beta_i \mu}{p_i} - \frac{\alpha_i}{\Sigma \, \alpha_k p_k} \right] , \tag{9.8}$$

where $\delta(.)$ is a function of a single variable.

Proof: We sketch the proof in four parts. (1) First, we show that (9.7) is the only non-degenerate admissible case of (9.2). Differentiating

$$\Sigma \, \psi^k [\log y_k - \log \Sigma \, w_k y_k] = 1$$

with respect to y_i, we find

$$\frac{\psi^{i\,\prime}}{y_i w_i} = \frac{\psi^{j\,\prime}}{y_j w_j} .$$

Differentiating with respect to y_s yields

$$\frac{\psi^{i\,\prime\prime}}{\psi^{i\,\prime}} = \frac{\psi^{j\,\prime\prime}}{\psi^{j\,\prime}} = c$$

where c is a constant. It is easy to show that $c = 0$ corresponds to a degenerate case. If $c \neq 0$, then

$$\psi^i(z_i) = \beta_i + \frac{\alpha_i}{c}$$

and the demand functions are given by (9.7). (2) There are no non-degenerate cases corresponding to (9.3). To show this we differentiate

$$\Sigma \, w_k y_k = T[\Sigma \, a_k \log y_k, \, \Sigma \, b_k \log y_k], \qquad \Sigma \, a_k = \Sigma \, b_k = 1$$

with respect to y_i

$$c_i y_i = a_i T_1 + b_i T_2 \qquad i = 1,\ldots,n$$

and observe that

$$\Sigma\, c_k y_k = T_1 + T_2 = T .$$

If two of the equations

$$c_i y_i = a_i T_1 + b_i T_2$$

are independent, we can solve for T_1 and T_2 in terms of the two corresponding y's. Hence, T depends only on those two y's, so this is a degenerate case. If all of the equations are linearly dependent, and some $c_i \neq 0$, we can take $c_1 \neq 0$ with no loss of generality. Then

$$a_i = \lambda a, \qquad b_i = \lambda b, \qquad c_i y_i = \lambda c_1 y_1 .$$

The last equation implies $\lambda = 0$, and hence $c_i = 0$, $i = 2,\ldots,n$. So $T(Y) = c_1 y_1$. (3) The only admissible case of (9.4) is a degenerate one in which only one is non-zero. (4) The non-degenerate admissible cases of (9.5) are those in which $c = 1$, which is (9.8).

We now consider two other price indexes which might be used to deflate p_i and μ in (9.1'): The Laspeyres and Paasche. Deflating by the Laspeyres index implies

$$T(P^t) = \frac{\Sigma\, h^k(P^b,\mu^b) p_k^t}{\mu^b} .$$

Since P^b and μ^b are constants, this is equivalent to requiring $T(P^t)$ to be a linear function of prices where the weights are proportional to reference period consumption.

This could happen in two ways. First, it might be that the demand functions are such that in every price expenditure situation the quantities demanded are proportional to the α's. That is,

$$\alpha_i = \lambda(P,\mu)h^i(P,\mu) \qquad i = 1,...,n \;.$$

Multiplying by p_i and summing over all goods, we find

$$\lambda(P,\mu) = \frac{\mu}{\Sigma \; \alpha_k p_k}$$

so the demand functions are of the homogeneous fixed coefficient form

$$h^i(P,\mu) = \frac{\alpha_i \mu}{\Sigma \; \alpha_k p_k} \;.$$

This proves that it is only in the homogeneous fixed coefficient case that the reference period consumption pattern is proportional to the α's for all price-expenditure situations. Second, even if every price-expenditure situation does not imply quantities demanded which are proportional to the α's , there are likely to be some price-expenditure situations in which proportionality holds. If, by a fortunate coincidence, the reference price-expenditure situation were one which implied proportionality, then deflation of price and expenditure by the Laspeyres index would be appropriate. There is little to say about the likelihood of this coincidence occurring, but even if it did occur, it is not clear how one would recognize it.

Deflating by the Paasche index is equivalent to setting

$$T(P^t) = \frac{\lambda(P^b)\mu^t}{\Sigma \; h^k(P^t,\mu^t)p_k^b}$$

since we require only proportionality. The factor of proportionality $\lambda(P^b)$ must be independent of both P^t and μ^t. As in the Laspeyres case, this may hold as an identity for all possible reference prices, or if it holds only for some reference prices, it may by coincidence hold for the particular situation we have observed. If it holds for all reference prices, then

$$T(P^t) \; \Sigma \; h^k(P^t,\mu^t)p_k^b = \lambda(P^b)\mu^t$$

is an identity in P^b. Differentiating with respect to p_i^b yields

$$T(P^t) \, h^i(P^t, \mu^t) = \lambda_i(P^b)\mu^t .$$

This implies that $\lambda_i(P^b)$ is a constant, say α_i, and so

$$h^i(P^t, \mu^t) = \frac{\alpha_i \mu^t}{T(P^t)} .$$

Multiplying by p_i and summing over all goods, we find

$$T(P^t) = \Sigma \, \alpha_k p_k^t$$

which implies that the demand functions are those of the homogeneous fixed coefficient case.

We summarize these results formally:

Theorem: If the demand functions are of the Fourgeaud-Nataf form (9.1)

$$h^i(P, \mu) = g^i \left[\frac{P_i}{T(P)}, \frac{\mu}{T(P)} \right], \qquad T(\lambda P) = \lambda T(P)$$

and if the index function $T(P)$ is proportional to either the Laspeyres or the Paasche index for all reference prices, then the direct utility function is of the homogeneous fixed coefficient form

$$U(X) = \min \left\{ \frac{x_i}{a_i} \right\}$$

and the demand functions are of the form

$$h^i(P,\mu) = \frac{a_i\mu}{\Sigma \, a_k p_k}.$$

This means that neither the Laspeyres nor the Paasche index has any special status in demand analysis. In particular, there is no presumption that deflating price and expenditure by the Laspeyres index is better than deflating by a weighted average of prices whose weights were chosen with the aid of a table of random digits. By the same token, there is no presumption that deflation of U.S. food prices and expenditure by a Laspeyres index based on U.S. consumption patterns is better, from a standpoint of demand analysis, than deflating by one whose weights reflect the consumption pattern of Outer Mongolia. The only way to find the appropriate weights is to estimate them along with the other unknown parameters.

Footnotes

[1] (added, 1982) For excellent recent surveys of the literature, see Diewert [1981, 1983]. A number of topics not treated here are discussed in my subsequent papers on cost-of-living indexes: Pollak, [1975a, 1975b, 1978, 1980, 1981, 1983], and Pollak and Wales [1979].

[2] (added, 1982) In the original version, μ was called "income"; somewhat inconsistently, I have retained such traditional phrases as "income elasticity" and "income-consumption" curve.

[3] (added, 1982) It is often, although not always, more convenient to denote a particular indifference curve by specifying a commodity vector X which lies on the indifference curve.

[4] (added, 1982) In the original 1971 version of this paper the expenditure function E(P,s) was called the "cost function" and denoted by C(P,s). The new terminology is especially convenient in the household production context where the term "cost function" is used in a different sense. See Pollak [1978]. It is often convenient to write the expenditure function as E(P,X), where X identifies the base indifference curve.

[5] (added, 1982). I now believe long-run pseudo preferences are inappropriate for welfare comparisons. See Pollak [1976].

[6] (added, 1982). In fact, these results do not depend on preferences satisfying the usual regularity conditions.

[7] (added, 1982). The procedure she describes in the published version of her paper, McElroy [1975] is more general than this sentence suggests.

[8] Some of the material in 6.1 and 6.2 is taken from Pollak [1971].

[9] The requirement $\Sigma\, a_k = 1$ is a normalization rule and involves no loss of generality.

[10] (added, 1982). This "generalized Cobb-Douglas" is a different form from Diewert's [1974, p.116].

[11] However, as Afriat [1972] points out, there are problems with regularity conditions in the homogeneous quadratic case.

[12] "Real wages" are sometimes used as a measure of quantity, but they are also used in empirical analysis of the supply or demand for labor. The appropriateness of using an index of real wages obtained in this way for empirical analysis is not explicitly discussed in this paper although the use of price indexes to deflate prices and expenditure in demand analysis is discussed in Section 9. The conclusion there is that such deflation is inappropriate, and it seems clear that deflation of money wages is no better.

[13] In Section 6 we used $\phi(P)$ to denote a function homogeneous of degree 1; in this section ϕ is not assumed to be homogeneous.

[14] There are further restrictions of the function T which we can safely ignore.

* This research was supported in part by the Bureau of Labor Statistics and the National Science Foundation. I am grateful to Irving B. Kravis and Karl Shell for helpful discussions, but neither they, the University of Pennsylvania, the Bureau of Labor Statistics, nor the National Science Foundation should be held responsible for the views expressed.

References

Afriat, S.N. "Theory of International Comparison of Real Income and Prices", In D.J. Daly, ed., *International Comparisons of Prices and Output*. New York: National Bureau of Economic Research, Columbia University Press, 1972.

Fisher, Irving. *The Making of Index Numbers*. Cambridge, Massachusetts: Houghton Mifflin Co., 1922.

Fisher, F.M. and K. Shell. "Taste and Quality Change in the Pure Theory of the True Cost-of-Living Index". In J.N. Wolfe, ed., *Value, Capital and Growth: Papers in Honour of Sir John Hicks*. Edinburgh: University of Edinburgh Press, 1968.

Fourgeaud, C. and A. Nataf. "Consommation en Prix et Revenu Reels et Theorie des Choix", *Econometrica*, Vol. 27, No. 3 (July 1959), pp. 329-354.

Gorman, W.M. "On a Class of Preference Fields", *Metroeconomica*, Vol. 13 (August 1961), pp. 53-56.

Klein, L.R. and H. Rubin. "A Constant-Utility Index of the Cost-of-Living", *Review of Economic Studies*, Vol. XV (2), No. 38 (1947-48), pp. 84-87.

Malmquist, S. "Index Numbers and Indifference Surfaces", *Trabajos de Estadistica* (1953), 4, pp. 209-242.

McElroy, Marjorie. "A 'Kinked' Linear Expenditure System: An Application to Rural South Vietnamese Households", mimeograph, 1969.

Pollak, Robert A. "Additive Utility Functions and Linear Engel Curves", *Review of Economic Studies*, Vol. 38, No. 4 (October 1971), pp. 401-414.

_____ "Generalized Separability", *Econometrica*, Vol. 40, No. 3 (May 1972), pp. 431-453.

_____ "Habit Formation and Dynamic Demand Functions", *Journal of Political Economy*, Vol. 78, No. 4 (July-August 1970), pp. 745-763.

Samuelson, Paul A. *Foundations of Economic Analysis*. Cambridge, Massachusetts: Harvard University Press, 1947.

Wegge, L.L. "The Demand Curves from a Quadratic Utility Indicator", *Review of Economic Studies*, Vol. XXXV (2), No. 102 (April 1968), pp. 209-224.

Wold, H.O.A. and L. Jureen. *Demand Analysis*. New York: John Wiley and Sons, 1953.

Additional References, 1982

Diewert, W.E., "Applications of Duality Theory", in M. Intriligator and D. Kendrick, eds., *Frontiers of Quantitative Economics II*, Amsterdam: North-Holland Publishing Co., 1974, pp. 106-171.

_____ "The Economic Theory of Index Numbers: A Survey", in A. Deaton, ed., *Essays in the Theory and Measurement of Consumer Behaviour in Honour of Sir Richard Stone*. London: Cambridge University Press, 1981, pp. 163-208.

_____ "The Theory of the Cost-of-Living Index and the Measurement of Welfare Change", *Price Level Measurement: Proceedings From a Conference Sponsored by Statistics Canada,* (1983).

McElroy, Marjorie, "A Spliced CES Expenditure System", *International Economic Review*, Vol. 16, No. 3 (October 1975), pp. 765-780.

Pollak, Robert A., "Subindexes of the Cost-of-Living Index", *International Economic Review*, Vol. 16, No. 1 (February 1975), pp. 135-150. (1975a)

_____ "The Intertemporal Cost-of-Living Index", *Annals of Economic and Social Measurement*, Vol. 4, No. 1, (Winter 1975), pp. 179-195. (1975b)

_____ "Habit Formation and Long-Run Utility Functions", *Journal of Economic Theory*, Vol. 13, No. 2 (October 1976), pp. 272-297.

_____ "Welfare Evaluation and the Cost-of-Living Index in the Household Production Model", *American Economic Review*, Vol. 68, No. 3, (June 1978), pp. 285-299.

_____ "Group Cost-of-Living Indexes", *American Economic Review*, Vol. 70, No. 2 (May 1980), pp. 273-278.

_____ "The Social Cost-of-Living Index", *Journal of Public Economics*, Vol. 15, No. 3, (June 1981), pp. 311-336.

_____ "The Treatment of 'Quality' in the Cost-of-Living Index", *Journal of Public Economics*, Vol. 20, No. 1, (February 1983), pp.25-53.

Pollak, Robert A., and Terence J. Wales, "Welfare Comparisons and Equivalence Scales", *American Economic Review*, Vol. 69, No. 2, (May 1979), pp. 216-221.

PRICE LEVEL MEASUREMENT – W.E. Diewert (Editor)
Canadian Government Publishing Centre /
Elsevier Science Publishers B.V. (North-Holland)

THE THEORY OF THE COST-OF-LIVING INDEX AND THE MEASUREMENT OF WELFARE CHANGE

W.E. Diewert*
Department of Economics
University of British Columbia

SUMMARY

The Consumer Price Index is often regarded as an approximation to a Cost-of-Living Index. This paper reviews the theoretical foundations of the Cost-of-Living Index and the closely related problems involved in measuring changes in economic welfare.

The Cost-of-Living Index for a single person is defined as the minimum cost of achieving a certain standard of living during a given period divided by the minimum cost of achieving the same standard of living during a base period. In order to numerically construct an individual's Cost-of-Living Index, it is necessary to know his or her preferences over economic goods. Since these preferences are essentially unobservable, it is necessary to construct approximations to the Cost-of-Living Index. This topic is discussed in Section 2 of the paper.

The remaining sections of the paper discuss a number of related topics, including: the closely related problems involved in measuring a group Cost-of-living index and changes in the welfare of a group, the fixed based versus the chain principle, the choice of a functional form for the Cost-of-Living Index, the treatment of durable goods, such as housing and the treatment of taxes and labour supply in a Cost-of-Living Index.

1. Introduction

As the title of the paper indicates, we will investigate the theoretical foundations of the cost-of-living index. This seems appropriate in a conference about the Consumer Price Index (CPI), since the CPI is now being used as a proxy for the cost-of-living index in indexing contracts and as an inflation measure.[1] We shall also discuss the closely related issues involved in measuring welfare changes, both for individual consumers and for groups of consumers.

The economic theory of the cost-of-living index for a single household is reviewed in Section 2.

In Section 3, we study the closely related problem of obtaining single household indicators of utility or real income change.

In Section 4, we discuss the costs and benefits of using the chain principle in the construction of index number formulae versus the fixed base principle.

In Sections 5 to 7, we discuss various concepts that have been proposed for group cost-of-living and welfare indexes.

In Sections 8 and 9, we outline the theory of subindexes of the cost-of-living index and the related idea of an intertemporal cost-of-living index.

In Section 10, we consider the problem of constructing price indexes that compare the level of prices in different locations.

Labour and durable goods in the cost-of-living index make their appearance in Sections 11 and 12 respectively.

Section 13 discusses the new goods problem and Section 14 concludes with some recommendations.

In a companion paper, Diewert [1983], we discuss price and output indexes from the viewpoint of producer theory. The reader may be aware of the old Hicks [1940; 1958; 1981] - Samuelson [1950; 1961] measurement of real income controversy; i.e., is there such an animal as real income, and if so, should it be measured from the consumer or producer point of view? From the consumer point of view, our conclusion is that real income is a very subjective animal and hence it probably does not exist, unless we are willing to give explicit numerical weights to the welfares of different household classes. On the producer side, the situation is more encouraging: although real output is not a useful concept, the closely related concept of total factor productivity does turn out to be useful. For the details of this "new" approach to measurement total factor productivity (which in fact was suggested many years ago by Hicks [1961] [1981; 192-3]), see Caves, Christensen and Diewert [1982b] and Diewert [1983].

Section 15 is an Appendix that collects proofs of new theorems.[2]

2. The Single Household Cost-of-Living Index

We assume that the household or individual has recurring preferences over combinations of N goods that may be represented by a utility function F where $u = F(x)$ is the utility level or standard of living that can be attained if the individual consumes the consumption vector $x \equiv (x_1, x_2,...,x_N)^T \geq 0_N{}^3$.

We assume that the utility function F satisfies Conditions I which are technical enough to relegate to a footnote.[4]

We shall assume that the consumer maximizes his utility function F(x) subject to a budget constraint of the form $p \cdot x = \Sigma_{n=1}^N p_n x_n \leq y$ where $p >> 0_N$ is a positive vector of commodity (rental) prices and $y > 0$ is expenditure on the N commodities.

The consumer's utility maximization problem can be decomposed into two stages. In the first stage, the consumer attempts to minimize the cost of achieving a given utility level, and in the second stage, he chooses the maximal utility level that is just consistent with his budget constraint.

The solution to the first-stage problem defines the consumer's *cost function* C: for u ≥ 0, $p >> 0_N$

$$C(u,p) \equiv \min_x \{p \cdot x : F(x) \geq u, \; x \geq 0_N\} \;. \tag{1}$$

Given that F satisfies Conditions I, C will satisfy Conditions II (which we relegate to another footnote[5]). Moreover, if we are given a cost function C satisfying Conditions II, C may be used in order to construct the underlying preference function F which will satisfy Conditions I.[6]

Our interest in C stems from the fact that it may be used to define the Konüs [1924] *cost-of-living index* P_K: for $p^0 >> 0_N$, $p^1 >> 0_N$ and $u > 0$ define

$$P_K(p^0, p^1, u) \equiv C(u, p^1)/C(u, p^0) \;. \tag{2}$$

Thus P_K depends on three variables: (i) p^0, a vector of period 0 or base period prices, (ii) p^1, a vector of period 1 or current period prices, and (iii) u, a number that indexes the reference indifference surface. Thus $P_K(p^0, p^1, u)$ is the minimum cost of achieving the standard of living indexed by u when the consumer faces period 1 prices p^1 **relative** to the minimum cost of achieving the same standard of living when the consumer faces period 0 prices p^0. If there is only one good, then it can be seen that $P_K(p_1^0, p_1^1, u) = p_1^1/p_1^0$ for all $u > 0$. In this case, there is obviously no index number problem.

In the general case when there is more than one good, the functional form for the cost-of-living index P_K obviously depends on the functional form for the consumer's cost function C, which in turn is determined by the form of the consumer's preference function F. Our fundamental problem is that we do not know what the functional forms for F or C and hence P_K are. Our primary task in this section will be to see if we can find adequate bounds or approximations to the true cost-of-living index P_K that depend only on observable market price and quantity data. However, before we turn to this primary task, we state a theorem which provides necessary and sufficient conditions for a given function $P(p^0, p^1, u)$ of $2N+1$ variables to be interpretable as a cost-of-living index.

Theorem 1

Let P be a function of $2N+1$ positive variables that satisfies the following properties: for all $u>0$, $p^0>>0_N$, $p^1>>0_N$, and $p^2>>0_N$, we have (i) $P(p^0, p^1, u) > 0$ (positivity), (ii) $P(p^0, p^1, u) = 1/P(p^1,p^0,u)$ (time reversal property), (iii) $P(p^0,p^2,u) = P(p^0,p^1,u)P(p^1,p^2,u)$ (circularity or transitivity) and (iv) for some $p^*>>0_N$, $C(u,p) \equiv uP(p^*,p,u)$ regarded as a function of u and p satisfies Conditions II for a cost function. Then P is the cost-of-living index that corresponds to the preferences that are dual to C; i.e., $P \equiv P_K$ satisfies(2). Moreover, C satisfies the following money metric[7] scaling of utility property:

$$C(u,p^*) = u \text{ for all } u > 0. \tag{3}$$

Conversely, given a cost function C satisfying Conditions II and the money metric property (3), then $P \equiv P_K$ defined by (2) satisfies properties (i) to (iv) listed above in the theorem.

The above theorem is very closely related to some results in Pollak [1983], who stressed that the mathematical properties of P_K are completely characterized by the mathematical properties of C.

The following theorem[8] provides observable bounds on $P_K(p^0,p^1,u)$.

Theorem 2

(Lerner [1935-36], Joseph [1935-36], Samuelson [1947; p.159]). P_K lies between the lowest price ratio and the highest price ratio for any reference indifference surface indexed by $u>0$; i.e.,

$$\min_i\{p^1_i/p^0_i: i=1,\ldots,N\} \le P_K(p^0,p^1,u)$$
$$\tag{4}$$
$$\le \max_i\{p^1_i/p^0_i: i=1,\ldots,N\}.$$

The above limits are wide, but they are not useless. For example, if prices vary in strict proportion so that $p^1 = \lambda p^0$ for some $\lambda > 0$, then the upper and lower limit in (4) is λ and hence $P_K(p^0, \lambda p^0, u) = \lambda$ also.

In order to make further progress, we shall assume cost minimizing behaviour on the part of the consumer during periods 0 and 1. We shall also assume that we can observe the consumer's quantity choices $x^0 > 0_N$ and $x^1 > 0_N$ made during periods 0 and 1 in addition to the corresponding price vectors $p^0 > > 0_N$ and $p^1 > > 0_N$. Thus we assume:

$$p^0 \cdot x^0 = C[F(x^0), p^0] \; ; \; p^1 \cdot x^1 = C[F(x^1), p^1]. \tag{5}$$

We have introduced the concept of the Konüs cost-of-living index $P_K(p^0, p^1, u)$ without saying much about the choice of the reference indifference surface indexed by u. It would appear that there are two natural choices for u: namely $F(x^0)$ or $F(x^1)$. Thus the *Laspeyres-Konüs cost-of-living index* is defined as:

$$P_K(p^0, p^1, F(x^0)) \equiv C[F(x^0), p^1]/C[F(x^0), p^0] \tag{6}$$

while the *Paasche-Konüs cost-of-living index* is defined as:

$$P_K(p^0, p^1, F(x^1)) \equiv C[F(x^1), p^1]/C[F(x^1), p^0]. \tag{7}$$

In order to understand why the indexes (6) and (7) are named after Laspeyres and Paasche it is first necessary to introduce the concept of a *mechanical price index formula*. This is simply a function P of the observable price and quantity vectors for the two periods, p^0, p^1, x^0, x^1, of known functional form. Two examples of such formulae are the *Laspeyres price index* P_L defined by

$$P_L(p^0, p^1, x^0, x^1) \equiv p^1 \cdot x^0/p^0 \cdot x^0 \tag{8}$$

and the *Paasche price index* P_P defined by

$$P_P(p^0,p^1,x^0,x^1) \equiv p^1 \cdot x^1 / p^0 \cdot x^1. \tag{9}$$

Irving Fisher [1922] gives hundreds of examples of mechanical price index formulae. The axiomatic characterization of these indexes may be found in Eichhorn [1976; 1978] and Eichhorn and Voeller [1976; 1983]. Note that these indexes are functions of 4N arguments whereas the price index that appeared in Theorem 1 had only $2N+1$ arguments.

The following theorem relates the (unobservable) Laspeyres-Konüs cost-of-living index defined by (6) to the (observable) Laspeyres price index defined by (8), and the Paasche-Konüs cost-of-living index defined by (7) to the Paasche price index defined by (9).

Theorem 3

(Konüs [1924; pp.17-19]): Assuming (5), cost minimizing behaviour during periods 0 and 1, we have:

$$P_K(p^0,p^1,F(x^0)) \le p^1 \cdot x^0 / p^0 \cdot x^0 \equiv P_L; \tag{10}$$

$$P_K(p^0,p^1, F(x^1)) \ge p^1 \cdot x^1 / p^0 \cdot x^1 \equiv P_P. \tag{11}$$

Corollary (Pollak [1983]):

$$\min_i \{p_i^1/p_i^0\} \le P_K(p^0,p^1,F(x^0)) \le P_L \equiv p^1 \cdot x^0 / p^0 \cdot x^0; \tag{12}$$

$$P_P \equiv p^1 \cdot x^1 / p^0 \cdot x^1 \le P_K(p^0,p^1, F(x^1)) \le \max_i \{p_i^1/p_i^0\}. \tag{13}$$

The above corollary follows combining Theorems 2 and 3. The Laspeyres-Konüs index $P_K(p^0,p^1,F(x^0)) \equiv C[F(x^0),p^1]/C[F(x^0),p^0] = p^1 \cdot x^{0*}/p^0 \cdot x^0$ is illustrated in Figure 1 in the two good case along with the bounds in (12). Note that x^{0*} is the solution to the problem of minimizing the cost of achieving the utility level $u^0 \equiv F(x^0)$ when the consumer is faced with period 1 prices p^1. Although $p^1 \cdot x^{0*}$ is not observable, the upper bound $p^1 \cdot x^0$

(see the dashed straight line through x^0) is observable. The lower bound to $p^1 \cdot x^{0*}$ is $p^1 \cdot \tilde{x}^0$, where $\tilde{x}^0$ is the point where the budget line $\{x: p^0 \cdot x = p^0 \cdot x^0\}$ intersects the x_1 axis, and it too is observable (see the dashed line through $\tilde{x}^0$). These upper and lower bounds correspond to the upper and lower bounds in (12). Analytically, the upper bound rests on the fact that the point x^0 is an **inner approximation** to the consumer's true indifference set $L(u^0) \equiv \{x: x \geq F(x^0) \equiv u^0\}$ while the set that is on or above the period 0 budget line $x: p^0 \cdot x \geq p^0 \cdot x^0$ intersected with the non-negative orthant $x: \{x: x \geq 0_N\}$ is an **outer approximation** to the true indifference set $L(u^0)$.

Figure 1.

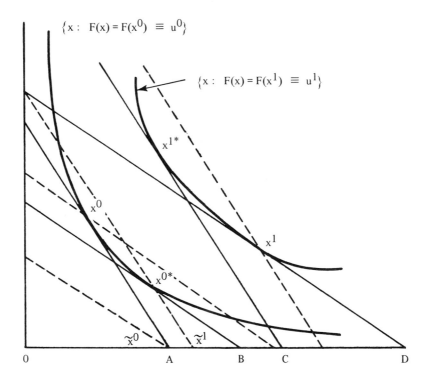

A similar analysis can be carried out for the Paasche-Konüs price index $P_K(p^0, p^1, F(x^1))$ $\equiv C[F(x^1), p^1]/C[F(x^1), p^0] = p^1 \cdot x^1 / p^0 \cdot x^{1*}$. From Figure 1, it can be seen that $p^0 \cdot x^1$ is an upper bound for $p^0 \cdot x^{1*}$ while $p^0 \cdot \tilde{x}^1$ is a lower bound for $p^0 \cdot x^{1*}$. These upper and lower bounds for $p^0 \cdot x^{1*}$ yield the lower and upper bounds in (13).

It can be seen from Figure 1, that the Laspeyres index P_L will be rather close to $P_K(p^0,p^1,F(x^0))$ provided that the indifference surface x: $F(x) = F(x^0)$ is not too linear around x^0. (The perfectly linear case corresponds to the perfect substitutes case). Similarly, the Paasche index P_P will be close to $P_K(p^0, p^1,F(x^1))$ provided that the indifference surface x: $\{F(x) = F(x^1)\}$ is not too linear around x^1. This is an encouraging observation, but it still does not tell us how close P_L is to $P_K(p^0,p^1,F(x^0))$ or how close P_P is to $P_K(p^0,p^1,F(x^1))$.

In order to make further progress, we may proceed in three directions: (i) introduce additional observations $(p^2,x^2),...,(p^T,x^T)$ and use the revealed preference techniques associated with Samuelson [1947] and Afriat [1967; 1972; 1979] in order to form non-parametric approximations to the consumer's preferences,[9] (ii) make specific functional form assumptions about F or C, or (iii) choose the reference utility level u that occurs in $P_K(p^0,p^1,u)$ in an empirically convenient manner. We will pursue only possibilities (ii) and (iii) in this paper.

It is an empirical fact that the Laspeyres and Paasche indexes, $P_L(p^0,p^1,x^0,x^1)$ and $P_P(p^0,p^1,x^0,x^1)$, are often rather close to each other numerically (we will return to this point in Section 4). Thus the following theorem is extremely important from a practical point of view.

Theorem 4

(Konüs [1924; pp.20-21]): Let the consumer's utility function F satisfy Conditions I and suppose that the observed data for periods 0 and 1, (p^0,x^0) and (p^1,x^1) respectively, satisfy the cost minimization assumptions (5). Then there exists a reference utility level u* that lies between[10] the base utility level $u^0 \equiv F(x^0)$ and the period 1 utility level $u^1 \equiv F(x^1)$ such that the consumer's true cost-of-living index for this reference utility level, $P_K(p^0,p^1,u^*)$, lies between $P_L \equiv p^1 \cdot x^0/p^0 \cdot x^0$ and $P_P \equiv p^1 \cdot x^1/p^0 \cdot x^1$; i.e., we have

$$P_P \leq P_K(p^0,p^1,u^*) \leq P_L \qquad \text{if } P_P \leq P_L \text{ or} \qquad (14)$$

$$P_L \leq P_K(p^0,p^1,u^*) \leq P_P \qquad \text{if } P_L \leq P_P . \qquad (15)$$

In most applications of index number theory, we would be quite happy if we knew $P_K(p^0,p^1,u^0)$ or $P_K(p^0,p^1,u^1)$ or $P_K(p^0,p^1,u^*)$ for some u^* between u^0 and u^1. Hence if P_L is numerically close to P_P, $P_P(p^0,p^1,u^*)$ will be squeezed in by these two numbers and we will have the consumer's true cost of living between periods 0 and 1 for all practical purposes.

It would be pleasant if we could extend Theorem 4 to conclude that there exists a u^* between u^0 and u^1 such that $P_K(p^0,p^1,u^*)$ equals some specific average of P_L and P_P, such as Irving Fisher's [1922] ideal index number formula P_2 which is defined as the geometric average of P_L and P_P; i.e.,

$$P_2(p^0,p^1,x^0,x^1) \equiv [p^1 \cdot x^0 / p^0 \cdot x^0]^{1/2} [p^1 \cdot x^1 / p^0 \cdot x^1]^{1/2}.$$

Unfortunately, we cannot draw such a conclusion in general. The problem is that $P_K(p^0,p^1,u)$ could be rather close to **either P_L or P_P** for all u between u^0 and u^1 and hence if P_L and P_P are rather different, then their geometric mean P_2 could lie above or below $P_K(p^0,p^1,u)$ for all u between u^0 and u^1. However, if P_L and P_P are "close" to each other, then there exists a u^* between u^0 and u^1 such that $P_K(p^0,p^1,u^*)$ is "close" to the Fisher index P_2. This provides a somewhat informal justification for the use of P_2 as an approximation to P_K.

A more formal justification for the use of P_2 that rests on a specific functional form for $C(u,p)$ is also possible. Let us suppose that $C(u,p) = uc^r(p)$ where c^r is defined by:

$$c^r(p) \equiv [\sum_{i=1}^{N} \sum_{j=1}^{N} b_{ij} p_i^{r/2} p_j^{r/2}]^{1/r} \text{ for } r \neq 0, \qquad (16)$$

where the parameters b_{ij} satisfy the restrictions $b_{ij} = b_{ji}$ for $1 \leq i < j \leq N$. The unit cost function c^r defined by (16) is Denny's [1974] quadratic mean of order r unit cost function which can provide a second order approximation to an arbitrary twice continuously differentiable unit cost function.[11]

Define the base period expenditure shares $s_i^0 \equiv p_i^0 x_i^0 / p^0 \cdot x^0$ and the period 1 expenditure shares $s_i^1 \equiv p_i^1 x_i^1 / p^1 \cdot x^1$ for $i = 1, \ldots, N$. For each $r \neq 0$, define the mechanical price index formula P_r by:

$$P_r(p^0, p^1, x^0, x^1) = [\sum_{i=1}^{N} s_i^0 (p_i^1 / p_i^0)^{r/2}]^{1/r} [\sum_{j=1}^{N} s_j^1 (p_j^1 / p_j^0)^{-r/2}]^{-1} r. \qquad (17)$$

It can be verified that when $r = 2$, P_r defined by (17) coincides with the Fisher index P_2 defined earlier.

The following theorem relates c^r defined by (16) to the price index formula P_r defined by (17).

Theorem 5

(Diewert [1976; p.133]): Suppose $C(u,p) = uc^r(p)$ and the observed data (p^0, x^0) and (p^1, x^1) for periods 0 and 1 satisfy the cost minimization assumption (5). Then for any reference utility level $u > 0$, we have

$$P_K(p^0, p^1, u) = c^r(p^1) / c^r(p^0) = P_r(p^0, p^1, x^0, x^1). \qquad (18)$$

Thus $P_K(p^0, p^1, u)$ may be precisely determined by using the mechanical index number formula P_r defined by (17). Diewert [1976] calls a price index formula $P(p^0, p^1, x^0, x^1)$ **superlative** if it correctly evaluates the Konüs price index $P_K(p^0, p^1, u) \equiv C(u, p^0) / C(u, p^1)$ for some cost function C of the form $C(u,p) = uc(p)$ where the unit cost function c can provide a second order approximation to an arbitrary unit cost function. Equations (16) to (18) show that the price indexes P_r are superlative for each $r \neq 0$.

Theorem 5 provides a strong economic justification for the use of the indexes P_r in empirical applications. In particular, we have obtained an economic justification for the use of the Fisher index P_2.

However, Theorem 5 is subject to a serious defect: namely, if $C(u,p) = u c^r(p)$, then the preference function F^r that is dual to this cost function is homothetic; in particular $F^r(x)$ is linearly homogeneous in x. Theorem 6 below is not subject to this defect.

First, let us define the **translog cost** function C^0 by:

$$\ln C^0(u,p) \equiv a_0 + \sum_{i=1}^{N} a_i \ln p_i + 1/2 \sum_{i=1}^{N} \sum_{j=1}^{N} a_{ij} \ln p_i \ln p_j + a_{00} \ln u$$

$$+ \sum_{i=1}^{N} a_{0i} \ln p_i \ln u + 1/2 a_{000}(\ln u)^2 \tag{19}$$

where the parameters a_{ij} satisfy the following restrictions.

$$\sum_{i=1}^{N} a_i = 1; \; a_{ij} = a_{ji}; \; \sum_{j=1}^{N} a_{ij} = 0 \text{ for } i=1,\ldots,N; \; \sum_{i=1}^{N} a_{0i} = 0. \tag{20}$$

Define the translog price index P_0^{12} by

$$P_0(p^0,p^1,x^0,x^1) \equiv \prod_{i=1}^{N} (p_i^1/p_i^0)^{(s_i^0 + s_i^1)/2} \tag{21}$$

Theorem 6

(Diewert [1976; p.122]: Suppose the consumer's cost function C equals the translog cost function C^0 defined by (19) and suppose the observed data (p^0,x^0) and (p^1,x^1) satisfy the cost minimization assumptions (5) where F is the utility function dual to C. Define $u^0 \equiv F(x^0)$, $u^1 \equiv F(x^1)$ and $u^* \equiv (u^0 u^1)^{1/2}$. Then the true cost of living index $P_K(p^0,p^1,u^*)$

evaluated at the intermediate utility level u^* may be calculated by evaluating the translog price index P_0; i.e., we have

$$P_K(p^0, p^1, u^*) \equiv C^0(u^*, p^1)/C^0(u^*, p^0) = P_0(p^0, p^1, x^0, x^1). \tag{22}$$

We note that the translog cost function C^0 defined by (19) can provide a second order approximation to an arbitrary twice continuously differentiable cost function, so that Theorem 6 is not restricted to the homothetic case. Thus Theorem 6 provides a very strong economic justification for the use of the translog price index P_0 as an approximation to the true cost of living $P_K(p^0, p^1, u^*)$.

Theorems 4 and 5 together provided a strong justification for the use of the Fisher index P_2 while Theorem 6 justified the use of the translog index P_0. Which one should we use? We will return to this question in Section 4, but first, it is useful to study the problem of measuring changes in the consumer's welfare (as opposed to the problem of measuring changes in the levels of prices that he faces).

3. Single Household Welfare Indicators

A first approach to measuring changes in the consumer's welfare would be to use the Konüs cost-of-living index $P_K(p^0, p^1, u)$ as a deflator for the consumer's expenditure ratio between the two periods, $p^1 \cdot x^1 / p^0 \cdot x^0$. Hence we define the *Pollak* [1971; p.64] *Implicit Quantity Index* $\tilde{Q}_K$ as

$$\tilde{Q}_K(p^0, p^1, x^0, x^1, u) \equiv p^1 \cdot x^1 / p^0 \cdot x^0 P_K(p^0, p^1, u). \tag{23}$$

If we make our usual cost minimization assumption (5), and if we use definition (2), we find that we can rewrite (23) as

$$\tilde{Q}_K(p^0, p^1, x^0, x^1, u) = \frac{C(u^1, p^1)}{C(u, p^1)} \cdot \frac{C(u, p^0)}{C(u^0, p^0)}, \tag{24}$$

where as usual $u^0 \equiv F(x^0)$ and $u^1 \equiv F(x^1)$.

There are two natural choices for the reference utility level u; namely, u^0 and u^1. Inserting these choices into (24) leads to the following formulae:

$$\tilde{Q}_K(p^0,p^1,x^0,x^1,u^0) = C(u^1,p^1)/C(u^0,p^1); \qquad (25)$$

$$\tilde{Q}_K(p^0,p^1,x^0,x^1,u^1) = C(u^1,p^0)/C(u^0,p^0). \qquad (26)$$

Diewert [1981: p.170] shows that $\tilde{Q}_K$ defined by (24) has the correct ordinal properties if the reference utility level u is chosen to be any level between u^0 and u^1 (including u^0 and u^1 as well); i.e., if $u^0 \le u \le u^1$ with at least one strict inequality, then $\tilde{Q}_K(p^0,p^1,x^0,x^1,u) > 1$ while if $u^0 \ge u \ge u^1$ with at least one strict inequality, then

$$\tilde{Q}_K(p^0,p^1,x^0,x^1,u) < 1.$$

The special cases of (23) defined by (25) and (26) may be illustrated with reference to Figure 1. The index defined by (25) is equal to the ratio OD/OB while the index defined by (26) is equal to the ratio OC/OA.

The special cases (25) and (26) are also special cases of another class of quantity indexes. For $x^0 > 0_N$, $x^1 > 0_N$ and $p >> 0_N$, define the *Allen* [1949; p.199] *Quantity Index*[13] as

$$Q_A(x^0,x^1,p) \equiv C[F(x^1), p]/C[F(x^0),p]. \qquad (27)$$

When $p = p^1$, (27) reduces to (25), and when $p = p^0$, (27) reduces to (26). The reader will also be able to verify that the implicit quantity index $\tilde{Q}_K$ defined by (24) is a product of two Allen quantity indexes.

The Pollak and Allen quantity indexes, $\tilde{Q}_K$ and Q_A, are studied in greater detail (and

bounds are derived) in Pollak [1971] and Diewert [1981]. We shall not dwell on their properties here since neither index is the most natural concept for a quantity index or a welfare
indicator. If $x^1 = \lambda x^0$ for some $\lambda > 0$, it would be desirable if our quantity index took on
the value λ. Neither $\tilde{Q}_K$ nor Q_A has this desirable homogeneity property. However, the
Malmquist [1953; 232] *Quantity Index* Q_M does have this homogeneity property. In order
to define Q_M, we must first define the **deflation** or distance function $D(u,x)$ that corresponds
to the consumer's utility function F. For $u > 0$ and $x >> 0_N$, define

$$D(u,x) \equiv \max_k \{k: F(x/k) \geq u, k > 0\}. \tag{28}$$

Thus $D(u,x^1)$ is the deflation factor k_1 say that will just reduce the vector x^1 propor-
tionately so that $F(x^1/k_1) = u$. If F satisfies Conditions I, then D will satisfy certain regularity conditions (Conditions III say) and a D satisfying these conditions will uniquely
characterize F.[14]

For $u > 0$, $x^0 >> 0_N$, $x^1 >> 0_N$, define the *Malmquist Quantity Index* Q_M as

$$Q_M(x^0,x^1,u) \equiv D(u,x^1)/D(u,x^0). \tag{29}$$

In general, the Malmquist quantity index $Q_M(x^0,x^1,u)$ will depend on the reference indifference surface indexed by u. As usual, two natural choices for u are $u^0 \equiv F(x^0)$ and
$u^1 \equiv F(x^1)$. Thus the *Laspeyres-Malmquist quantity index* is defined as

$$Q_M(x^0,x^1,u^0) \equiv D(u^0,x^1)/D(u^0,x^0)$$

$$= D(u^0,x^1) \tag{30}$$

where (30) follows since $D(u^0,x^0) = 1$. The *Paasche-Malmquist quantity index* is defined as

$$Q_M(x^0,x^1,u^1) \equiv D(u^1,x^1)/D(u^1,x^0)$$

$$= 1/D(u^1,x^0) \tag{31}$$

since $D(u^1, x^1) = 1$.

Geometric interpretations for the general Malmquist index (29) and the two special indexes (30) and (31) may be obtained from Figure 2 for the case of two goods. An observed quantity vector x^0 is the point A on the u^0 indifference curve while the other observed quantity vector x^1 is the point F on the u^1 indifference curve. The reference indifference curve indexed by u is indicated by the dashed indifference curve.

Figure 2.

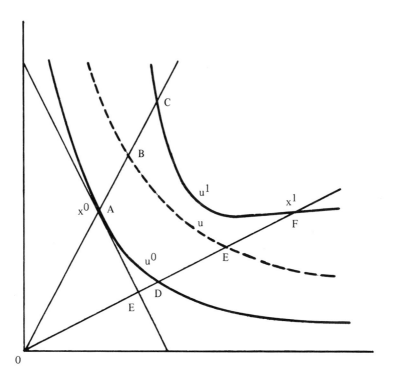

The reader can confirm that $Q_M(x^0, x^1, u)$ defined by (29) is equal to $[OF/OE]/[OA/OB]$, that $Q_M(x^0, x^1, u^0) = OF/OD$ and that $Q_M(x^0, x^1, u^1) = OC/OA$.

Note that the assumption of cost minimizing behaviour is **not** required in order to define the Malmquist quantity index.

From Figure 2, we see that as long as the reference indifference surface indexed by u remains between u^0 and u^1, the Malmquist index $Q_M(x^0, x^1, u)$ will correctly indicate whether welfare has increased or decreased going from x^0 to x^1. This property holds in general as the following result indicates.[15]

Theorem 7

(Diewert [1981; p.174]): If F satisfies Conditions I, $x^0 > > 0_N$, $x^1 > > 0_N$ and u is between $F(x^0)$ and $F(x^1)$, then (i) $Q_M(x^0, x^1, u) = 1$ if $F(x^0) = F(x^1)$, (ii) $Q_M(x^0, x^1, u) < 1$ if $F(x^0) > F(x^1)$, and (iii) $Q_M(x^0, x^1, u) < 1$ if $F(x^0) < F(x^1)$.

Comparing definitions (2), which defined P_K as a ratio of cost functions, with definition (29), which defined Q_M as a ratio of distance functions, it could be conjectured that Q_M will satisfy more or less the same mathematical properties as P_K, except that the role of prices and quantities is interchanged. This conjecture is correct. Thus we could write down a Q_M version of Theorem 1, where Q_M replaces P_K, x's replace p's, D replaces C, and Conditions III for distance functions D replace Conditions II for cost functions C. Similarly, Theorem 8 below is an exact analogue to Theorem 2.

Theorem 8

(Diewert [1981; p.175]): If F satisfies Conditions I and $x^0 > > 0_N$, $x^1 > > 0_N$, $u > 0$, then

$$\min_i \{x_i^1/x_i^0\} \leq Q_M(x^0, x^1, u) \leq \max_i \{x_i^1/x_i^0\}.$$

The above theorem provides a start to the problem of providing observable points for the essentially unobservable index Q_M. The above theorem did not require the assumption of cost minimizing behaviour on the part of the consumer. The following theorem does.

Theorem 9

(Malmquist [1953; p.231]): Suppose F satisfies Conditions I and (5) holds. Define u^0 $\equiv F(x^0)$ and $u^1 \equiv F(x^1)$. Then

$$Q_M(x^0,x^1,u^0) \leq p^0 \cdot x^1/p^0 \cdot x^0 \equiv Q_L(p^0,p^1,x^0,x^1), \text{ and} \qquad (33)$$

$$Q_M(x^0,x^1,u^1) \geq p^1 \cdot x^1/p^1 \cdot x^0 \equiv Q_P(p^0,p^1,x^0,x^1). \qquad (34)$$

Leontief [1936; pp.58-59] illustrated the above bounds in the two good case. Note that the right-hand side of (33) defines the *Laspeyres quantity index* Q_L and the left-hand side of (33) is the Laspeyres-Malmquist quantity index defined earlier by (30). Thus Q_L is an observable upper bound to the essentially unobservable Malmquist index $Q_M(x^0,x^1,u^0)$. In Figure 2, $Q_M(x^0,x^1,u^0) = 0F/0D \leq 0F/0G = p^0 \cdot x^1/p^0 \cdot x^0 = Q_L$.

The right-hand side of (34) defines the *Paasche quantity index* Q_P and the left-hand side of (34) is the Paasche-Malmquist quantity index defined earlier by (31). Thus Q_P is an observable lower bound to $Q_M(x^0,x^1,u^1)$.

Corollary

$$\min_i \{x_i^1/x_i^0\} \leq Q_M(x^0,x^1,F(x^0)) \leq Q_L \equiv p^0 \cdot x^1/p^0 \cdot x^0; \qquad (35)$$

$$p^1 \cdot x^1/p^1 \cdot x^0 \equiv Q_P \leq Q_M(x^0,x^1,F(x^1)) \leq \max_i \{x_i^1/x_i^0\} \qquad (36)$$

Note that $Q_M(p^0,p^1,x^0,x^1)$ $P_P(p^0,p^1,x^0,x^1) = p^1 \cdot x^1/p^0 \cdot x^0 = Q_P(p^0,p^1,x^0,x^1)$ $P_L(p^0,p^1,x^0,x^1)$. Thus if the Paasche and Laspeyres price indexes P_P and P_L are numerically close, then the Paasche and Laspeyres price indexes P_P and P_L are numerically close, then the Paasche and Laspeyres quantity indexes Q_P and Q_L will also be close. Thus the upper bound in (35) will often be close to the lower bound in (36). Hence the following theorem is extremely useful.

Theorem 10

(Diewert [1981; p.176]): Suppose F satisfies Conditions I and the cost minimization assumption (5) holds. Then there exists a reference utility level u^* between $u^0 \equiv F(x^0)$ and $u^1 \equiv F(x^1)$ such that the Malmquist quantity index $Q_M(x^0, x^1, u^*)$ for this reference utility level u^* lies between Q_L and Q_P.

Theorem 10 is a counterpart to Theorem 4 (and in fact is proved in the same manner). Thus if Q_L and Q_P are "close" to each other, we may take an average of them (such as Irving Fisher's ideal quantity index $Q_2 \equiv (Q_L Q_P)^{1/2}$) and obtain a "close" approximation to the unobservable Malmquist quantity index $Q_M(x^0, x^1, u^*)$ where u^* is some utility level between u^0 and u^1.

Note that Theorem 10 made no assumptions about the shape of the consumer's indifference surfaces (other than our usual general regularity conditions on the consumer's utility function F). However our choice of the reference utility level u^* was somewhat limited.

In order to make further progress, it is necessary to make specific functional form assumptions. Thus let $f^r(x)$ be the quadratic mean of order r utility function for $r \neq 0$, defined in a manner analogous to the definition of $c^r(p)$ (recall (16)) and define the quadratic mean of order r mechanical quantity index formula Q_r by

$$Q_r(p^0, p^1, x^0, x^1) \equiv P_r(x^0, x^1, p^0, p^1) \tag{37}$$

where P_r was defined by (17). Note that we have interchanged the role of prices and quantities in the right-hand side of (37). It may be verified that when $r = 2$, Q_2 defined by (37) reduces to the Fisher quantity index, $(Q_P Q_L)^{1/2}$.

The following theorem is the quantity counterpart to Theorem 5.

Theorem 11

(Diewert [1976; p.132]): Suppose $F = f^r$ for some $r \neq 0$ and the data (p^0, x^0), (p^1, x^1) satisfy the cost minimization assumption (5). Then for any reference utility level $u > 0$, we have

$$Q_M(x^0, x^1, u) = f^r(x^1)/f^r(x^0) = Q_r(p^0, p^1, x^0, x^1). \tag{38}$$

Thus $Q_M(x^0, x^1, u)$ may be precisely determined by using the mechanical index number formula Q_r defined by (37), provided that $F = f^r$. Since f^r can provide a second order approximation to an arbitrary twice continuously differentiable homogeneous utility function, Diewert [1976] calls Q_r a **superlative** quantity index number formula.

Although Theorem 11 provides a strong economic justification for the use of the superlative indexes Q_r, the result is subject to the same defect that occurred in Theorem 5; namely, the preferences that correspond to f^r are homothetic. Theorem 12 below is not subject to this limitation.

First, we define the translog distance function $D^0(u, x)$ by setting $\ln D^0(u, x)$ equal to the right-hand side of (19) where x_i replaces p_i. Define the *translog quantity index* Q_0 by

$$\ln Q_0(p^0, p^1, x^0, x^1) = \sum_{i=1}^{N} \frac{1}{2} \left\{ \frac{p_i^0 x_i^0}{p^0 \cdot x^0} + \frac{p_i^1 x_i^1}{p^1 \cdot x^1} \right\} \ln(x_i^1/x_i^0). \tag{39}$$

Theorem 12

(Diewert [1976; p.123]): Suppose the consumer's distance function D equals the translog distance function D^0 defined above. Let F be the corresponding utility function and C the corresponding cost function. Let the observed data (p^0, x^0), (p^1, x^1) satisfy the cost minimization assumptions (5). Define $u^0 \equiv F(x^0)$, $u^1 \equiv F(x^1)$ and $u^* \equiv (u^0 u^1)^{1/2}$. Then the Malmquist quantity index $Q_M(x^0, x^1, u^*)$ evaluated at the intermediate utility level u^* is precisely equal to the translog quantity index Q_0; i.e.,

$$Q_M(x^0, x^1, u^*) = Q_0(p^0, p^1, x^0, x^1).$$

Since the translog distance function D^0 can provide a second order approximation to an arbitrary twice continuously differentiable distance function, Theorem 12 provides a very strong economic justification for the use of the translog quantity index Q_0 as an approximation to the Malmquist quantity index $Q_M(x^0, x^1, u^*)$.

4. Fixed Base versus Chain Indexes

In Section 2, we found a family of mechanical index number formulae, $P_r(p^0, p^1, x^0, x^1)$ for each number r, which had *a priori* good properties from an economic point of view. To each such P_r, there corresponds an implicit quantity index $\tilde{Q}_r$ that may be defined as follows:

$$\tilde{Q}_r(p^0, p^1, x^0, x^1) \equiv p^1 \cdot x^1 / p^0 \cdot x^0 \; P_r(p^0, p^1, x^0, x^1). \tag{40}$$

In Section 3, we found a family of mechanical quantity indexes, $Q_r(p^0, p^1, x^0, x^1)$ for each number r, which had *a priori* good properties from an economic point of view. To each such Q_r, there corresponds an implicit price index $\tilde{P}_r$ defined by:

$$\tilde{P}_r(p^0, p^1, x^0, x^1) \equiv p^1 \cdot x^1 / p^0 \cdot x^0 \; Q_r(p^0, p^1, x^0, x^1). \tag{41}$$

Our problem now is that we have too many index number formulae that have desirable properties. Which price index formula should we choose from the double infinity of candidates of the form P_r or $\tilde{P}_s$ for some r or s? The following theorem leads to an answer to this question.

Theorem 13

(Diewert [1978]): The functions $P_r(p^0, p^1, x^0, x^1)$ and $\tilde{P}_s(p^0, p^1, x^0, x^1)$ differentially approximate each other to the second order around any point where $p^0 = p^1 >> 0_N$ and $x^0 = x^1 >> 0_N$. A similar statement holds for the quantity indexes Q_s and $\tilde{Q}_r$.

Thus we have for all r and s;

$$P_r(p^0,p^0,x^0,x^0) = \tilde{P}_s(p^0,p^0,x^0,x^0) \tag{42}$$

$$\nabla P_r(p^0,p^0,x^0,x^0) = \nabla \tilde{P}_s(p^0,p^0,x^0,x^0) \text{ and} \tag{43}$$

$$\nabla^2 P_r(p^0,p^0,x^0,x^0) = \nabla^2 \tilde{P}_s(p^0,p^0,x^0,x^0) \tag{44}$$

where ∇P_r stands for the 4N dimensional vector of first order partial derivatives of the function P_r and $\nabla^2 P_r$ is the 4N by 4N matrix of second order partials of P_r.

Hence for "normal" time series data, the indexes $P_r(p^0,p^1,x^0,x^1)$ and $\tilde{P}_s(p^0,p^1,x^0,x^1)$ **will all give the same answer** to a very high degree of approximation. Some empirical evidence on this issue is presented in Diewert [1978; 1983b] and in Généreux [1983]. Thus the answer to the question posed before Theorem 13 is: it does not matter very much which formula we use.

Diewert [1978] also shows that the Paasche and Laspeyres indexes that figured prominently in sections 2 and 3 approximate the superlative indexes to the first order, i.e., for $p^0 >> 0_N$ and $x^0 >> 0_N$, we have

$$P_r(p^0,p^0,x^0,x^0) = P_L(p^0,p^0,x^0,x^0) = P_P(p^0,p^0,x^0,x^0) \text{ and} \tag{45}$$

$$\nabla P_r(p^0,p^0,x^0,x^0) = \nabla P_L(p^0,p^0,x^0,x^0) = \nabla P_P(p^0,p^0,x^0,x^0) \tag{46}$$

but we do not obtain the equality of the second order partial derivatives as we did in (44).

It is worth emphasizing that the above results hold **without** the assumption of optimizing behaviour on the part of consumers; i.e., they are theorems in numerical analysis rather then economics.[16]

Theorems 4 and 10 above showed that it is very useful to have the Paasche and Laspeyres indexes close to each other since this will lead to a very close approximation to the Konüs cost-of-living index $P_K(p^0,p^1,u^*)$ and to the Malmquist quantity index $Q_M(x^0,x^1,u^*)$. This implies that we should use the **chain** principle for constructing indexes rather than the **fixed base** principle. The difference between the two principles may be illustrated for the case of three observations. In the fixed base principle, given a mechanical index number formula P, the aggregate level of prices for the three periods would be

$$1, \; P(p^0,p^1,x^0,x^1) \text{ and } P(p^0,p^2,x^0,x^2). \tag{47}$$

Using the chain principle, the aggregate price level for the three periods would be

$$1, \; P(p^0,p^1,x^0,x^1) \text{ and } P(p^0,p^1,x^0,x^1)P(p^1,p^2,x^1,x^2).$$

Using the chain principle, variations between prices and quantities tend to be smaller and hence the Paasche, Laspeyres and all of the superlative indexes will approximate each other much more closely than is the case when we use a fixed base. This is borne out in empirical computations.[17]

If we had to choose **a** particular index number formula for a price index from the families P_r and $\tilde{P}_s$, we could perhaps narrow the choice to three indexes: (i) P_0, because of Theorem 6, (ii) $\tilde{P}_0$, because of Theorem 12, or (iii) P_2, because it lies between the Paasche and Laspeyres bounds.[18]

Of course, there is a high **empirical** cost associated with the use of a superlative index number formula as opposed to using a simple fixed based Laspeyres index of the form $p^t \cdot x^0 / p^0 \cdot x^0$. The advantage of this latter index is that it requires quantity information for only the base period. Its disadvantage is that it will probably[19] not approximate the consumer's true cost-of-living index $P_K(p^0,p^t,u^0)$ very closely as p^t moves further away from p^0.

Having discussed price and welfare indexes for an individual consumer or household at great length, it is time now to turn to the construction of group indexes.

5. The Democratic Consumer Price Index

Assume that there are H distinct households, where household h has utility function $F^h(x)$ satisfying Conditions I noted in Section 2 with a dual cost function $C^h(u_h, p)$.

There are many possible ways for constructing a group cost-of-living index. Our first method will be to construct a simple average of the individual household Konüs cost-of-living indexes $C^h(u_h, p^1)/C^h(u_h, p^0)$. This is what Prais [1959] and Muellbauer [1974] call a *democratic price index*. Letting $u \equiv (u_1, ..., u_H)^T$ denote a vector of reference utility levels, we define the *democratic cost-of-living index* P_D as

$$P_D(p^0, p^1, u) \equiv \sum_{h=1}^{H} \frac{1}{H} \frac{C^h(u_h, p^1)}{C^h(u_h, p^0)}. \qquad (49)$$

Let $p^0 >> 0_N$ and $p^1 >> 0_N$ be the period 0 and 1 price vectors as usual, let $x_h^t \equiv (x_{1h}^t, x_{2h}^t, ..., x_{Nh}^t)^T$ be consumer h's observed quantity vector for period t, t = 0, 1 and for h = 1,..., H, let $u_h^0 \equiv F^h(x_h^0)$ and $u_h^1 \equiv F^h(x_h^1)$ be the period 0 and 1 utility levels attained by household h, and define the utility vectors $u^1 \equiv (u_1^1, ..., u_H^1)^T$ and $u^0 \equiv (u_1^0, ..., u_H^0)^T$. Assume cost minimizing behaviour for each household during both periods; i.e.,

$$p^0 \cdot x_h^0 = C^h(u_h^0, p^0) = C^h[F^h(x_h^0), p^0] \qquad \text{for } h = 1,...,H \qquad (50)$$

and

$$p^1 \cdot x_h^1 = C^h(u_h^1, p^1) = C^h[F^h(x_h^1), p^1] \qquad \text{for } h = 1,...,H.$$

Then we may immediately apply Theorems 2 and 3 in order to obtain the following bounds on the Laspeyres-Democratic index $P_D(p^0, p^1, u^0)$ and on the Paasche-Democratic index $P_D(p^0, p^1, u^1)$:

$$\min_i\{p_i^1/p_i^0\} \le P_D(p^0,p^1,u^0) \le \sum_{h=1}^{H} \frac{1}{H} \frac{p^1 \cdot x_h^0}{p^0 \cdot x_h^0}; \qquad (51)$$

$$\sum_{h=1}^{H} \frac{1}{H} \frac{p^1 \cdot x_h^1}{p^0 \cdot x_h^1} \le P_D(p^0,p^1,u^1) \le \max_i\{p_i^1/p_i^0\}. \qquad (52)$$

Note that the right-hand side of (51) is an arithmetic average of the individual household Laspeyres price indexes, $\Sigma_h\, P_L(p^0,p^1,x_h^0,x_h^1)/H$, which we denote by $\bar{P}_L$. The left-hand side of (52) is an arithmetic average of the individual Paasche indexes, $\Sigma_h P_P(p^0,p^1,x_h^0,x_h^1)/H$, which we denote by $\bar{P}_P$.

The following theorem provides a group counterpart to Theorem 4 (and in fact is proven in the same manner).

Theorem 14

Let each consumer's utility function F^h satisfy Conditions I. Suppose that the observed data for periods 0 and 1, (p^0,x_h^0) and (p^1,x_h^1) for $h = 1,...H$, satisfy the cost minimization assumption (50). Then there exists a reference utility vector $u^* \equiv (u_1^*,...,u_H^*)$ such that each component u_h^* lies between u_h^0 and u_h^1 and the Prais-Muellbauer group cost-of-living index evaluated at this reference utility vector $P_D(p^0,p^1,u^*)$, lies between $\bar{P}_L$ and $\bar{P}_P$.

For typical data, we would expect $\bar{P}_L$ and $\bar{P}_L$ to be very close to each other (closer than the individual indexes P_L^h and P_P^h since $\bar{P}_L$ and $\bar{P}_P$ are **averages** of the individual P^h, and P_P^h), so $P_D(p^0,p^1,u^*)$ may be very closely approximated by either $\bar{P}_L$ or $\bar{P}_P$.

The practical difficulty with using the democratic price index $P_D(p^0,p^1,u^*)$ as a general measure of inflation between periods 0 and 1 is that $\bar{P}_L$ and $\bar{P}_P$ can only be constructed if we have individual household data. The group index defined in the following section has bounds that can be constructed from aggregate data.

6. The Plutocratic Cost-of-Living Index

Pollak [1981; p.328] defines what he calls a Scitovsky-Laspeyres cost-of-living index for a group of consumers as the ratio of the total expenditure required to enable each household to attain its reference indifference curve at period 1 prices to that required at period 0 prices. This same concept of a group cost-of-living index was suggested by Prais [1959] in less precise language. Prais referred to his concept as a *plutocratic price index*. Making the same assumptions and using the same notation as in the previous section, this Prais-Pollak *plutocratic cost-of-living index* may be defined as:

$$P_{PP}(p^0,p^1,u) \equiv \sum_{h=1}^{H} C^h(u_h,p^1) \Big/ \sum_{h=1}^{H} C^h(u_h,p^0) \tag{53}$$

$$= \sum_{h=1}^{H} s^h(u,p^0)\, C^h(u_h,p^1)/C^h(u_h,p^0) \tag{54}$$

where the consumer h share of total expenditure at reference utility levels u prices p is defined as

$$s^h(u,p) \equiv C^h(u_h,p) \Big/ \sum_{h=1}^{H} C^h(u_h,p) \qquad \text{for } h=1,...,H. \tag{55}$$

The *Laspeyres-Pollak cost-of-living index* is defined as $P_{PP}(p^0,p^1,u^0)$ while the *Paasche-Pollak cost-of-living index* is defined as $P_{PP}(p^0,p^1,u^1)$. Note that each household is given the same weight $1/H$ in the democratic price index defined by (49), while household h is given the expenditure share weight $s^h(u,p^0)$ in the plutocratic price index defined by (54).

It is convenient to define aggregate consumption vectors $\bar{x}^t$ as follows:

$$\bar{x}^0 \equiv \sum_{h=1}^{H} x_h^0 ; \qquad \bar{x}^1 \equiv \sum_{h=1}^{H} x_h^1 . \tag{56}$$

Armed with the above definitions, we may now prove the usual Paasche and Laspeyres bounding theorem.

Theorem 15

Suppose each consumer's utility function F^h satisfies Conditions I and suppose the cost minimization assumptions (5) hold. Then

$$\min_i\{p_i^1/p_i^0\} \leq P_{PP}(p^0,p^1,u^0) \leq p^1 \cdot \bar{x}^0/p^0 \cdot \bar{x}^0 \equiv P_L \qquad (57)$$

$$P_P \equiv p^1 \cdot \bar{x}^1/p^0 \cdot \bar{x}^1 \leq P_{PP}(p^0,p^1,u^1) \leq \max_i\{p_i^1/p_i^0\}. \qquad (58)$$

Note that P_L in (57) is the usual Laspeyres price index involving the **aggregate** base period consumption vector $\bar{x}^0$, while P_P in (58) is the usual Paasche price index involving the **aggregate** period 1 consumption vector $\bar{x}^1$. The consumer price index constructed by statistical agencies is usually an approximation to P_L. All that is required to construct P_L is: the current vector of prices p^1, the base price vector p^0, and the base period aggregate consumption vector $\bar{x}^0 \equiv \Sigma_h x_h^0$.

Usually, the upper bound P_L in (57) will be close to $P_{PP}(p^0,p^1,u^0)$ while the lower bound P_P in (58) will be close to $P_{PP}(p^0,p^1,u^1)$ and to P_L.

Theorem 16

Under the conditions of Theorem 15, there exists a reference utility vector $u^* \equiv (u_1^*,...,u_H^*)$ such that each component u_h^* lies between u_h^0 and u_h^1 and the Prais-Pollak group cost-of-living index evaluated at this reference utility vector, $P_{PP}(p^0,p^1,u^*)$, lies between $P_L \equiv p^1 \cdot \bar{x}^0/p^0 \cdot \bar{x}^0$ and $P_P \equiv p^1 \cdot \bar{x}^1/p^0 \cdot \bar{x}^1$.

As was the case for the Prais-Muellbauer index P_D, we would expect P_L and P_P to be very close to each other provided that p^0 and p^1 are not "too" different, and hence $P_{PP}(p^0,p^1,u^*)$ may be closely approximated by either P_L or P_P (or say by the Fisher ideal $P_2 \equiv P_L P_P^{1/2}$) under these circumstances.

We have noted that the bounds for the Pollak index, P_L and P_P, may be computed provided only that we have aggregate data on prices and quantities, while the bounds for the Muellbauer index, $\overline{P}_L$ and $\overline{P}_P$, require detailed household data for their computation. Are there other important differences between the two concepts of the group price index?

An important conceptual difference emerges if we compare the formula (49) for P_D and the formula (54) for P_{PP}. In (49), each household's individual Konüs cost-of-living index, $C^h(u_h, p^1)/C^h(u_h, p^0)$, is given the same weight, $1/H$. In (54), when $u = u^0$, household h's Laspeyres-Konüs cost-of-living index, $C^h(u_h^0, p^1)/C^h(u_h^0, p^0)$, is given the weight $s^h(u^0, p^0)$, which is household h's share of total expenditure during period 0. Thus high expenditure households will tend to get weighted more heavily than low expenditure households in the construction of the plutocratic index. This is why Prais and Muellbauer call the index (49) a democratic index, since it weights each household's cost-of-living index equally.

Which concept of a group price index should be used as a measure of average inflation between periods 0 and 1? Unless we have quantity data by household class, we cannot evaluate the bounds for the democratic index, so in this case we are stuck with the plutocratic index. If we do have detailed data by household class, then instead of calculating bounds for the democratic index, we should calculate the usual Paasche and Laspeyres bounds for the **individual** households true cost-of-living indexes, $C^h(u_h, p^1)/C^h(u_h, p^0)$.

Deaton and Muellbauer [1980; p.178] note that the simplification of working with a single price index can be very dangerous. They cite the Great Bengal Famine of 1943 when between three and five million people died of starvation. An average price index was not very relevant to the problems of low-income households under those circumstances.

Statistics Canada [1982; p.89] has taken the first step in the direction of providing consumer price indexes that are household specific in that they now construct an experimental CPI for low-income families. They are to be commended for this effort and they should be given additional resources to construct additional price indexes by household class.

We turn now to a brief discussion of the mirror image to the problem of constructing group price indexes - the problem of constructing aggregate welfare indexes.

7. Social Welfare Indexes

In Section 3, we found that the Malmquist quantity index, $Q_M(x^0, x^1, u) \equiv D(u, x^1)/D(u, x^0)$, was a very satisfactory quantity index or welfare indicator for a single household. Let us suppose that household h's preferences may be represented by the deflation function $D^h(u_h, x_h)$ that is dual to the utility function $F^h(x_h)$, where F^h satisfies Conditions I as usual.

Define the N by H matrix of period 0 (1) consumer choices by X^0 (X^1): i.e.,

$$X^0 \equiv [x_1^0, x_2^0, ..., x_H^0]; \qquad X^1 \equiv [x_1^1, x_2^1, ..., x_H^1] \qquad (59)$$

where $x_h^t \equiv [x_{1h}^t, x_{2h}^t, ..., x_{Nh}^t]^T$ is consumer h's observed consumption vector during period t for $t = 0$, 1 and $h = 1, 2, ..., H$.

It is natural to try and aggregate individual welfare changes into a single scalar measure of overall welfare change. A simple *social index of welfare change* that respects individual preferences is

$$W(X^0, X^1, u) \equiv \sum_{h=1}^{H} \beta_h(u) \, D^h(u_h, x_h^1)/D(u_h, x_h^0) \qquad (60)$$

where $u \equiv (u_1, u_2, ..., u_H)^T$ is a vector of individual reference utility levels and the weight function β_h satisfies the following restriction:

$$\sum_{h=1}^{H} \beta_h(u) = 1 \qquad \text{for any vector of reference utilities } u. \qquad (61)$$

The reader will note that our indicator of social welfare change defined by (60) is a quantity counterpart to the Pollak cost-of-living index defined by (54) (which is an indicator of price change).

Our reason for imposing the restriction (61) is the following: if $x_h^1 = x_h^0$ for $h = 1, \ldots, H$ so that there is no change in consumption between periods 0 and 1, then we want our indicator of social welfare change (or our aggregate quantity index) to indicate that there has been no change; i.e., we want $W(X^0, X^0, u) = 1$. Hence we must have (61).

In general, the weights $\beta_h(u)$ do not have to be positive or even non-negative. However, if the weights are non-negative, then we obtain the following bounds for W applying Theorem 8:

$$\min_{i,h}\{x_{ih}^1/x_{ih}^0\} \leq W(X^0, X^1, u) \leq \max_{i,h}\{x_{ih}^1/x_{ih}^0\} . \tag{62}$$

As usual, we define the *Laspeyres Social Welfare Indicator* $W(X^0, X^1, u^0)$ by selecting our reference utility vector to be $u^0 \equiv [F^1(x_1^0), \ldots, F^H(x_H^0)] \equiv [u_1^0, \ldots, u_H^0]$ and the *Paasche Social Welfare indicator* $W(X^0, X^1, u^1)$ by selecting the reference utility vector to be $u^1 \equiv [u_1^1, u_2^1, \ldots, u_H^1]$.

The following theorem provides a social welfare counterpart to Theorem 9.

Theorem 17

Let each household utility function F^h satisfy Conditions I and suppose the cost minimization assumptions (50) hold. Define the utility weights $\beta_h^0 \equiv \beta(u^0)$ and the utility weights $\beta_h^1 \equiv \beta_h(u^1)$ for $h = 1, 2, \ldots, H$. Then if the weights β_h^0 and β_h^1 are non-negative,

$$W(X^0, X^1, u^0) \leq \sum_{h=1}^{H} \beta_h^0 \, p^0 \cdot x_h^1 / p^0 \cdot x_h^0 \equiv \sum_{h=1}^{H} \beta_h^0 \, Q_L^h \tag{63}$$

and

$$\sum_{h=1}^{H} \beta_h^1 \, Q_P^h \equiv \sum_{h=1}^{H} \beta_h^1 \, p^1 \cdot x_h^1 / p^1 \cdot x_h^0 \leq W(X^0, X^1, u^1). \tag{64}$$

Thus the Laspeyres social welfare indicator $W(X^0, X^1, u^0)$ is bounded from above by a weighted average of the individual household Laspeyres quantity indexes Q_L^h, and the Paasche social welfare indicator $W(X^0, X^1, u^1)$ is bounded from below by a weighted average of the individual household Paasche quantity indexes Q_P^h.

The following theorem provides a social welfare analogue to Theorem 10.

Theorem 18

Suppose that the hypotheses of Theorem 17 are satisfied. Suppose also that the household weighting functions $\beta_h(u)$ are continuous as u varies linearly between u^0 and u^1. Then there exists a reference utility vector $u^* = (1-\lambda^*)u^0 + \lambda^* u^1$ for some λ^* between 0 and 1 such that the social welfare change indicator $W(X^0, X^1, u^*)$ lies between $\Sigma_{h=1}^H \beta_h^0 Q_L^h$ (the upper bound in (63)) and $\Sigma_{h=1}^H \beta_h^1 Q_P^h$ the lower bound in (64)).

Corollary

Suppose $\beta_h^0 = p^0 \cdot x_h^0 / \Sigma_{k=1}^H p^0 \cdot x_k^1$ (the share of household h in base period expenditure) and $\beta_h^0 = p^1 \cdot x_h^0 / \Sigma_{k=1}^H p^1 \cdot x_k^0$ (the share of household h in expenditure using period 1 prices and period 0 quantities) for $h = 1, ..., H$. Then the social welfare change indicator $W(X^0, X^1, u^*)$ defined in the theorem lies between the aggregate Laspeyres and Paasche quantity indexes, $p^0 \cdot (\Sigma x_h^1) / p^0 \cdot (\Sigma x_h^0)$ and $p^1 \cdot (\Sigma x_h^1) / p^1 \cdot (\Sigma x_h^0)$, respectively.

Thus if the aggregate Paasche and Laspeyres quantity indexes are close to each other, an average of them such as $Q_2 \equiv (Q_L Q_P)^{1/2}$ (Fisher's ideal quantity index), will yield a close approximation to the welfare change indicator $W(X^0, X^1, u^*)$ described in the above corollary. Thus we have provided a justification of sorts for Pigou's [1920; p.84] cautious recommendation of the Fisher quantity index as an indicator of aggregate welfare change.

However, I do not think that the corollary to Theorem 18 should be taken too seriously. We need very special welfare weights $\beta_h(u)$ in order to obtain the corollary. These special weights need not correspond to anybody's idea of a just society.

The paragraph above illustrates a problem with the social welfare approach: it is difficult to come to a consensus on what the weights $\beta_h(u)$ should be. Hence perhaps we should concentrate on the calculation of welfare change by household class rather than attempting to construct somewhat arbitrary measures of aggregate welfare change. For alternative attempts to construct aggregate cost-of-living indexes and aggregate measures of welfare change, see Blackorby and Donaldson [1983] and Jorgenson and Slesnick [1983].

We leave the world of group indexes in the following sections in order to focus on some special problems that are associated with single household price and quantity indexes.

8. The Theory of Subindexes

Pollak [1975a] noted that in many instances, economists are interested in *subindexes* of the cost-of-living index; i.e., in indexes that do not cover the whole spectrum of consumer goods, but only selected subsets. He notes several interesting classes of subindexes: (1) a food index say in the context of a complete cost-of-living index defined over all consumer goods, (2) a one-period index in the context of a multi-period world, (3) a consumption goods subindex in the context of a consumer choice model that included not only the consumption decision, but also the labour supply decision, (4) a consumption goods subindex in the context of a model where the consumer has preferences defined over not only consumer goods but also environmental variables (such as pollution) and also public goods (such as roads and parks).

Pollak's discussion summarized above indicated that the concept of a subindex in the cost-of-living index is not without applications. We must now face up to two problems: (i) how do we define a subindex rigorously, and (ii) how may we combine subindexes in order to form an approximation to the true overall cost-of-living index.[20]

Some new notation and a new concept are required. As usual, think of x as being the consumer's overall consumption vector. We now partition the vector x into M subvectors (of varying dimension) which we denote by $(x_1, x_2, \ldots, x_M) \equiv x$ (so x_m is the mth subvector). Partition the overall price vector p in an analogous manner. We may now define

Pollak's [1969][21] *conditional expenditure or cost function* for the mth subgroup of goods as:

$$C^m(u, p_m, x_1, x_2, \ldots, x_{m-1}, x_{m+1}, x_{m+2}, \ldots, x_M) \tag{65}$$

$$\equiv \min_{x_m} \{p_m \cdot x_m : F(x) \geq u\} \qquad m = 1, 2, \ldots, M,$$

where u is the consumer's reference utility level, F is his overall utility function satisfying the usual conditions, and the overall consumption vector is $x \equiv (x_1, x_2, \ldots, x_{m-1}, x_m, x_{m+1}, \ldots, x_M)$. Thus C^m defined by (65) is the minimum group m cost of achieving utility level u, given that the consumer has available x_1 units of group 1 goods, x_2 units of group 2 goods,..., x_{m-1} units of group m-1 goods, x_{m+1} units of group m + 1 goods,..., and x_M units of group M goods. In order to save space, we shall write C^m defined by (65) as $C^m(u, p_m, x)$ where x is the entire consumption vector, but it should be understood that $C^m(u, p_m, x)$ is constant with respect to variations in the x_m components of x; i.e., $C^m(u, p_m, x)$ does not depend on the x_m components of X.

The regularity properties of C^m with respect to the vector of price variables p_m are exactly the same properties that were given in Conditions II for the vector of price variables p.[22] Thus we may define Pollak's [1975; 147] *Subindex of the Cost of Living Index* for Group m goods as:

$$P^m(p_m^0, p_m^1, u, x) \equiv C^m(u, p_m^1, x)/C^m(u, p_m^0, x); \qquad m = 1, \ldots, M, \tag{66}$$

where $p^0 \equiv (p_1^0, p_2^0, \ldots, p_M^0)$ is the base period price vector (remember each p_m^0 is a vector pertaining to group m goods), $p^1 \equiv (p_1^1, p_2^1, \ldots, p_M^1)$ is the period 1 price vector, u is a reference utility level, and $x \equiv (x_1, x_2, \ldots x_M)$ is a reference quantity vector.

Since Theorem 2 depended only on the regularity properties of C(u,p) with respect to the price vector p and since $C^m(u, p_m, x)$ satisfies these same regularity properties, it can be seen that P^m satisfies the usual Lerner-Joseph-Samuelson bounds for any reference vector (u,x):

$$\min_i \left\{ p^1_{mi}/p^0_{mi} \right\} \leq P^m(p^0_m, p^1_m, u, x) \leq \max_i \left\{ p^1_{mi}/p^0_{mi} \right\}$$

where the index i runs through the components of the group m price vectors.

As usual, we can pick out particular reference vectors of interest. Define the *Laspeyres-Pollak Group* m *Subindex* as $P^m(p^0_m, p^1_m, u^0, x^0)$ where x^0 is the consumer's period 0 choice vector and $u^0 \equiv F(x^0)$ is his period 0 utility level. Define the *Paasche-Pollak Group* m *Subindex* as $P^m(p^0_m, p^1_m, u^1, x^1)$ where x^1 is the consumer's period 1 choice vector and $u^1 \equiv F(x^1)$ is his period 1 utility level. Assuming optimizing behaviour during the two periods, we may derive the following subindex counterpart to Theorem 3.

Theorem 19

Assuming (5), the mth subindex P^m defined by (66) satisfies the following inequalities when $(u,x) = (u^0,x^0)$ and (u^1,x^1) respectively:

$$P^m(p^0_m, p^1_m, u^0, x^0) \leq P^m_L \equiv p^1_m \cdot x^0_m / p^0_m \cdot x^0_m; \qquad m = 1,\ldots,M \text{ and} \qquad (67)$$

$$P^m(p^0_m, p^1_m, u^1, x^1) \geq P^m_P \equiv p^1_m \cdot x^1_m / p^0_m \cdot x^1_m; \qquad m = 1,\ldots,M. \qquad (68)$$

Thus we obtain the usual Laspeyres and Paasche bounds for the subindexes, and hence we may obtain the following adaptation of Theorem 4.

Theorem 20

Assume that F satisfies Conditions I, and assume that (5) holds; i.e., that there is overall utility maximizing behaviour during the two periods. Then there exists a reference utility-consumption vector $(u^{m*}, x^{m*}) \equiv (1-\lambda^*_m)(u^0,x^0) + \lambda^*_m(u^1,x^1)$ where $0 < \lambda^*_m < 1$ and $u^0 \equiv F(x^0)$, $u^1 \equiv F(x^1)$ such that the mth Pollak subindex evaluated at this reference vector, $P^m(p^0_m, p^1_m, u^{m*}, x^{m*})$,[23] lies between the group m Laspeyres index P^m_L defined in (67) and the group m Paasche index P_P^m defined in (68).

Thus under normal circumstances when P_L^m is close to P_P^m, we may obtain a rather good estimate of the subindex $P^m(p_m^0, p_m^1, u^{m*}, x^{m*})$.

How may we combine the subindexes in order to obtain an estimate of the overall cost-of-living $P_K(p^0, p^1, u)$?

Consider the Laspeyres-Pollak subindexes $P^m(p_m^0, p_m^1, u^0, x^0)$. A natural way of weighting these subindexes to form an overall cost-of-living index would be to use the base period expenditure share $p_m^0 \cdot x_m^0 / p^0 \cdot x^0$ to weight the mth subindex $P^m(p_m^0, p_m^1, u^0, x^0)$. Call the resulting overall cost-of-living index P(0). Thus

$$P(0) \equiv \Sigma_{m=1}^M \ (p_m^0 \cdot x_m^0 / p^0 \cdot x^0) P^m(p_m^0, p_m^1, u^0, x^0)$$

$$= \Sigma_{m=1}^M \ (p_m^0 \cdot x_m^0 / p^0 \cdot x^0) \ C^m(u^0, p_m^1, x^0) / C^m(u^0, p_m^0, x^0)$$

$$= \Sigma_{m=1}^M \ (p_m^0 \cdot x_m^0 / p^0 \cdot x^0) \ C^m(u^0, p_m^1, x^0) / p_m^0 \cdot x_m^0$$

using (A6) in the appendix

$$\leq p^1 \cdot x^0 / p^0 \cdot x^0 \equiv P_L \text{ using (A8) in the appendix.} \qquad (69)$$

Thus the aggregate two-stage index P(0) is bounded from above by the aggregate Laspeyres price index P_L. Note that if P^m were replaced by the Laspeyres group m subindex $p_m^1 \cdot x_m^0 / p_m^0 \cdot x_m^0$, then (69) would collapse to P_L, the aggregate Laspeyres price index.

Consider now the Paasche-Pollak subindexes $P^m(p_m^0, p_m^1, u^1, x^1)$. A natural way of weighting these indexes to form an overall cost-of-living index would be to use period 1 expenditure shares $p_m^1 \cdot x_m^1 / p^1 \cdot x^1$ as weights for the subindexes $P^m(p_m^0, p_m^1, u^1, x^1)$. However, instead of forming a simple weighted average, we shall form a harmonic mean:

$$P(1) \equiv [\Sigma^M_{m=1}(p^1_m \cdot x^1_m/p^1 \cdot x^1) \; [P^m(p^0_m,p^1_m,u^1,x^1)]^{-1}]^{-1}$$

$$= [\Sigma^M_{m=1}P^1_m \cdot x^1_m \; C^m(u^1,p^0_m,x^1)/p^1 \cdot x^1 C^m(u^1,p^1_m,x^1)]^{-1}$$

$$= [\Sigma^M_{m=1} \; C^m(u^1,p^0_m,x^1)/p^1 \cdot x^1]^{-1} \qquad \text{using (A7)}$$

$$\geq p^1 \cdot x^1/p^0 \cdot x^1 \equiv P_P \qquad\qquad \text{using (A9).} \qquad\qquad (70)$$

Thus the two stage-index $P(1)$ is bounded from below by the aggregate Paasche index P_P. Note that if P^m were replaced by the Paasche group M subindex $p^1_m \cdot x^1_m/p^0_m \cdot x^1_m$, then (70) would collapse to P_P, the aggregate Paasche price index. Thus the overall Paasche price index P_P may be calculated as a weighted harmonic average of Paasche subindexes.

For each λ between 0 and 1, define the intermediate aggregate two-stage index $P(\lambda)$ by

$$P(\lambda) \equiv [\; \overset{M}{\underset{m=1}{\Sigma}} \; [(1-\lambda)s^0_m + \lambda s^1_m] \; [P^m(p^0_m,p^1_m,(1-\lambda)u^0 + \lambda u^1,(1-\lambda)x^0 + \lambda x^1)]^r]^{1/r} \qquad (71)$$

where $s^0_m \equiv p^0_m \cdot x^0_m/p^0 \cdot x^0$, $s^1_m \equiv p^1_m \cdot x^1_m/p^1 \cdot x^1$ and the exponent r that appears in (71) is defined by $r \equiv 1-2\lambda$.

It can be verified that $P(0)$ and $P(1)$ defined by (71) coincide with the $P(0)$ and $P(1)$ defined in (69) and (70) and moreover, $P(\lambda)$ is continuous for $0 \leq \lambda \leq 1$. Thus we may apply the usual Konüs [1924] proof (see the proof of Theorem 14) and conclude that there exists a λ^* between 0 and 1 such that

$$P_P \leq P(\lambda^*) \leq P_L \text{ or } P_L \leq P(\lambda^*) \leq P_P. \qquad\qquad (72)$$

Thus $P(\lambda^*)$, a weighted average of the subindexes $P^m(p^0_m,p^1_m,(1-\lambda^*) u^0 + \lambda^* u^1,$ $(1-\lambda^*)x^0 + \lambda^* x^1)$ for $m=1,...,M$ has the usual overall Paasche and Laspeyres bounds and hence will be "close" to the one-stage Konüs cost-of-living index, $P_K(p^0,p^1,u^*)$, whose existence was given in Theorem 4 above.

Diewert [1978] showed that certain index number formulae, such as the superlative mechanical price index formulae $P_r(p^0, p^1, x^0, x^1)$ and $\tilde{P}_s(p^0, p^1, x^0, x^1)$ defined by (17) and (41), had a second order consistency in aggregation property.[24] Diewert's proof was highly technical and lacked economic motivation. Perhaps the results in this section cast some light on why "good" index number formulae may be expected to aggregate up to the "right" aggregate value when we use a "good" weighting scheme: the Paasche and Laspeyres bounds occur in an appropriate two-stage procedure as well as in the usual single-stage procedure for constructing a true cost-of-living index.

9. An Intertemporal Cost-of-Living Index

Pollak [1975a] developed the theory of subindexes and in Pollak [1975b], he attempted to apply his general theory of subindexes to the intertemporal context. However, in this section, we shall attempt to show that the general theory of subindexes cannot be very readily applied to the intertemporal context.

We begin with a single consumer who has a horizon that extends over $T + 1$ Hicksian [1946] periods when we first observe his behaviour in period 0. Let his utility function be given by $F(x_0, x_1, \ldots, x_T)$ where x_t is the period t vector of purchases that the consumer plans to make in period t. Assume for simplicity that the consumer can borrow or lend a dollar from period 0 to period 1 at the interest rate r_1^0 and he **expects** the interest rate from period t to $t + 1$ to be r_{t+1}^0 for $t = 1, 2, \ldots, T-1$. Define the sequence of period 0 expected discount factors to be $\delta_1^0 \equiv 1/(1 + r_1^0)$, $\delta_2^0 \equiv \delta_1^0/(1 + r_2^0), \ldots, \delta_T^0 \equiv (\delta_{T-1}^0/1 + r_T^0)$. Suppose that from the vantage point of period 0, the consumer **expects** the spot price vector $p_t^0 > 0_N$ to prevail during period t for $t = 1, 2, \ldots, T$. The price vector $p_0^0 > 0_N$ is the **observable** vector of market prices prevailing during period 0. Let p^0 be the vector of discounted expected prices

$$p^0 \equiv (p_0^0, \delta_1^0 p_1^0, \delta_2^0 p_2^0, \ldots, \delta_T^0 p_T^0).$$

We assume that $x^0 \equiv (x_0^0, x_1^0, \ldots, x_T^0)$ solves the period 0 expected utility maximization problem:

$$\max_x \left\{ F(x): p^0 \cdot x \le p^0 \cdot x^0 \equiv W^0 \right\} \tag{73}$$

where $W^0 > 0$ is the consumer's initial wealth. Letting C be the cost function: we have the following equation (where $\delta_0^0 \equiv 1$):

$$W^0 = C[F(x^0), p^0]$$

$$\equiv \min_x \left\{ p^0 \cdot x: F(x) \ge F(x^0) \right\}$$

$$= \min_{x_0, \ldots, x_T} \left\{ \sum_{t=0}^{T} \delta_t^0 p_t^0 \cdot x_t : F(x_0, \ldots, x_T) \ge F(x_0^0, \ldots, x_T^0) \right\}$$

$$= \sum_{t=0}^{T} \delta_t^0 p_t^0 \cdot x_t^0$$

$$= \sum_{t=0}^{T} C^t[F(x^0), \delta_t^0 p_t^0, x^0] \qquad \text{using (A6)}$$

$$= C^0[F(x^0), p_0^0, x^0] + \sum_{t=1}^{T} \delta_t^0 C^t[F(x^0), p_t^0, x^0] \tag{74}$$

where (74) follows from the homogeneity properties of the conditional cost functions C^t which are defined in a manner analogous to C^m in (65). In period 0, we may observe the consumer's initial wealth W^0, the vector of prevailing market prices p_0^0, the consumer's vector of actual period 0 purchases x_0^0, his savings $W^0 - p_0^0 \cdot x_0^0$, and perhaps the *ex ante* period 0 interest rate r_1^0 (and the corresponding discount rate $\delta_1^0 \equiv 1/(1 + r_1^0)$). *All of the other variables are unobservable.*

Now suppose that we can observe what happens during period 1. There will be new expectations about nominal interest rates r_t^1 for $t = 2,\ldots,T$ and a new sequence of discount factors $\delta_t^1 \equiv \delta_{t-1}^1/(1 + r_t^1) > 0$, the consumer will have a new initial wealth W^1, there will be a new spot price vector for goods p_1^1 (which may or may not be equal to the consumer's period 0 expectation about this vector of spot prices p_1^0), and the consumer will have new expectations about future spot prices, $p_2^1, p_3^1, \ldots, p_T^1$. We now assume that $x_1^1, x_2^1, \ldots, x_T^1$ solves the following period 1 expected utility maximization problem:

$$\max_{x_1,\ldots,x_T} \{F(x_0^0, x_1, \ldots, x_T): p_1^1 \cdot x_1 + \delta_2^1 p_2^1 \cdot x_2 + \cdots + \delta_T^1 \, p_T^1 \cdot x_T \tag{75}$$

$$\leq p_1^1 \cdot x_1^1 + \delta_2^1 p_2^1 \cdot x_2^1 + \cdots + \delta_T^1 \, p_T^1 \cdot x_T^1 = W^1\}.$$

Note that the period 0 decision vector x_0^0 appears in (75), since we are stuck with it in period 1. Define $x^1 \equiv (x_0^0, x_1^1, \ldots, x_T^1)$. Proceeding in a manner analogous to our derivation of (74), we find that

$$W^1 \equiv C^1[F(x^1), p_1^1, x^1] + \sum_{t=2}^{T} \delta_t^1 \, C^t[F(x^1), p_t^1, x^1]. \tag{76}$$

In period 1, we may observe the consumer's period 1 wealth W^1, the period 1 vector of actual market prices p_1^1, the consumer's purchases of goods during the period x_1^1, his saving $W^1 - p_1^1 \cdot x_1^1$, and the *ex post* rate of return that was earned going from period 0 to period 1, r_1^1 (remember the corresponding *ex ante* expected rate of return was r_1^0).

Using the techniques outlined in the previous section, we may establish the following observable bounds for two intertemporal subindexes:

$$\min_i \{p_{1i}^1/p_{0i}^0\} \leq \frac{C^0[F(x^0), \, p_1^1, x^0]}{C^0[F(x^0), p_0^0, x^0]} \leq p_1^1 \cdot x_0^0/p_0^0 \cdot x_0^0 \equiv P_L;$$

$$P_P \equiv p_1^1 \cdot x_1^1/p_0^0 \cdot x_1^1 \leq \frac{C^1[F(x^1), \, p_1^1, \, x^1]}{C^1[F(x^1), \, p_0^0, \, x^1]} \leq \max_i \{p_{1i}^1/p_{0i}^0\}.$$

However, the bounds established above do not allow us to answer any really interesting questions.

Although we cannot compute a complete intertemporal cost-of-living index in general, it is possible to make a comparison of the expected welfare of the consumer during periods 0 and 1. Recall that the *ex post* period 0 rate of return was defined to be r_1^1. Define the *ex post* discount rate $\delta_1^1 \equiv 1/(1+r_1^1)$. Then we may rewrite the period 1 expected utility maximization problem (75) in the following manner:

$$\max_{x_1,\ldots,x_T} \{F(x_0^0,x_1,\ldots,x_T): p_0^0 \cdot x_0^0 + \delta_1^1 p_1^1 \cdot x_1 + \delta_1^1 \delta_2^1 p_2^1 \cdot x_2 + \ldots \tag{77}$$

$$+ \delta_1^1 \delta_T^1 p_T^1 \cdot x_T \leq p_0^0 \cdot x_0^0 + \delta_1^1 w^1\}.$$

The objective functions in (75) and (77) are the same and the constraint in (77) is obtained by multiplying both sides of the constraint in (75) by δ_1^1 and then adding the constant $p_0^0 \cdot x_0^0$ to both sides.

The consumer's period 0 expected utility maximization problem (73) may be rewritten as (78) when we fix $x^0 = x_0^0$:

$$\max_{x_1,\ldots,x_T} \{F(x_0^0,x_1,\ldots,x_T): p_0^0 \cdot x_0^0 + \delta_1^0 p_1^0 \cdot x_1 + \delta_2^0 p_2^0 \cdot x_2 \tag{78}$$

$$+\ldots+ \delta_T^0 p_T^0 \cdot x_T \leq w^0\}.$$

If the consumer's period 0 expectations equal his period 1 expectations about future prices and interest rates and if the expected *ex ante* first period rate of return r_1^0 equals the *ex post* rate of return r_1^1, then it can be verified that the prices appearing in the constraint of (77) are identical to the prices appearing in the constraint of (78). Under these conditions, we could say that the choice set of the consumer (and hence his welfare) has increased going from period 0 to period 1 if

$$p_0^0 \cdot x_0^0 + \delta_1^1 w^1 > w^0 \tag{79}$$

where $\delta_1^1 \equiv 1/(1 + r_1^1)$ and r_1^1 is the *ex post* one-period rate of return on assets going from period 0 to period 1.

The criterion for an expected welfare increase (or for an increase in real wealth) may be rewritten as

$$w^1 > (w^0 - p_0^0 \cdot x_0^0)(1 + r_1^1)$$

$$= \text{(first period savings)}(1 + \text{ex post rate of return}).$$

This criterion for an increase in real wealth is due to Hicks [1946; p.175]. However, its validity does require the assumption of constant expectations, an assumption that is unlikely to be fulfilled under present economic conditions.

The problem of measuring welfare changes in a general Hicksian intertemporal choice model seems to be inherently difficult. Virtually all of the index number techniques that we have surveyed and developed in this paper rely on the twin assumptions of optimizing behaviour on the part of the consumer and observability of market prices and the consumer's quantity choices (or we require assumptions about the constancy of expectations that are unlikely to be met in practice). Hence in order to apply the traditional theory of index numbers in the intertemporal context, it appears to be necessary to place *a priori* restrictive assumptions on the form of the intertemporal utility function $F(x_0, x_1, ..., x_T)$ such as intertemporal additivity;[25] i.e., $F(x_0, x_1, ..., x_T) = \Sigma_{t=0}^{T} f(x_t)$. If we do this, then we may apply traditional index number theory in order to measure changes in the one-period utility function $f(x_t)$; i.e., we could approximate the change in $f(x_1^1)/f(x_0^0)$ using the period 0 and 1 price and quantity data in the usual manner.

10. Spatial Cost-of-Living Indexes

The basic problem to be considered in this section is the problem of comparing the level of prices in different cities or localities. The problem is isomorphic to the usual problem of making international comparisons.[26]

If we are willing to assume that a certain class of consumers in one location (location 0 say) has the same one-period preferences as another class of consumers in another location (location 1 say) and if there are no significant differences in environmental (non-market) variables in the two locations, then we may simply apply the theory outlined in Sections 2-4 above, where the superscripts 0 and 1 will now refer to locations. However, there were two rather big "ifs" in the previous sentence.

We may relax the restrictiveness of the second "if" by using a Pollak subindex (recall Section 8) of the form

$$C^1[F(x^0),p_1^1,x_2^0]/C^1[F(x^0),p_1^0,x_2^0] \equiv P^1(p_1^0,p_1^1,F(x^0),x_2^0) \text{ and} \tag{80}$$

$$C^1[F(x^1),p_1^1,x_2^1]/C^1[F(x^1),p_1^0,x_2^1] \equiv P^1(p_1^0,p_1^1,F(x^1),x_2^1] \tag{81}$$

where P^1 is a Pollak subindex of the cost of living over market goods (recall (66)), p_1^0 and p_1^1 are vectors of observed market prices in locations 0 and 1 respectively, $F(x) = F(x_1,x_1)$ is the consumer's utility function defined over combinations of market goods x_1 and non-market locational amenities x_2, x_1^0 and x_1^1 are the observed market choice vectors for consumers 0 and 1 respectively, x_2^0 is the amenity vector in location 0, x_2^1 is the amenity vector in location 1, $x^0 \equiv (x_1^0,x_2^0)$ and $x^1 \equiv (x_1^1,x_2^1)$.

The bounds derived in Section 8 (see (67) and (68)) are applicable to the subindexes defined in (80) and (81) under the usual optimizing behaviour assumptions:

$$\min_i \{p_{1i}^1/p_{1i}^0\} \leq P^1(p_1^0,p_1^1,F(x^0),x_2^0) \leq p_1^1 \cdot x_1^0/p_1^0 \cdot x_1^0 \equiv P_L \text{ and} \tag{82}$$

$$P_P \equiv p_1^1 \cdot x_1^1/p_1^0 \cdot x_1^1 \leq P^1(p_1^0,p_1^1,F(x^1),x_2^1) \leq \max_i \{p_{1i}^1/p_{1i}^0\}. \tag{83}$$

Theorem 19 may be applied in the present context as well.

The assumption that the consumers in the two locations have the same preferences may also be relaxed; see Caves, Christensen and Diewert [1982b; p.1410] and Denny and Fuss [1983] for various "translog approaches".

11. Leisure and Labour Supply in the Cost-of-Living Index

Consider a household that can supply various kinds of labour service. It is natural to assume that the household has one-period preferences defined over various combinations of market goods $x \equiv (x_1,...,x_N) \geq 0_N$ and labour supplies $y \equiv (y_1,...,y_M) \leq 0_M$ where $y_m \leq 0$ is the **negative** of the number of hours of the mth type of work supplied by the household. The preferences are summarized by the utility function $F(x,y)$.[27] Suppose now that the household faces the positive commodity price vector $p >> 0_N$ and the positive (after tax marginal) wage vector $w >> 0_M$. Then we may define the household's cost function C in the usual manner:

$$C(u,p,w) \equiv \min_{x,y} \{p \cdot x + w \cdot y : F(x,y) \geq u\}. \qquad (84)$$

Given period 0 and 1 price vectors, (p^0,w^0) and (p^1,w^1), we may be tempted to define the Konüs cost-of-living index in the usual manner:

$$P_K(p^0,w^0,p^1,w^1,u) \equiv C(u,p^1,w^1)/C(u,p^0,w^0). \qquad (85)$$

If we could be assured that $C(u,p,w) > 0$ for $u > 0$, and $p >> 0_N$, and $w >> 0_M$, then there would be no problem with definition (85), and in fact we could derive the usual bounds that we derived in Section 2. However, $C[F(x^0,y^0), p^0,w^0] > 0$ corresponds to a situation where the value of household consumption in period 0, $p^0 \cdot x^0$ exceeds the value of household labour supply, $-w^0 \cdot y^0$. There is no reason for this to be the case. Worse yet, the two values could coincide (i.e., we could have $p^0 \cdot x^0 + w^0 \cdot y^0 = 0$) in which case $P_K(p^0,w^0,p^1,w^1, F(x^1,y^1))$ becomes undefined (since we are dividing by zero in definition (85)).

We could attempt to avoid these problems by assuming that the household has preferences defined over different combinations of market goods x and leisure, where the leisure vector is defined by $b + y \geq 0_M$ and $y \leq 0_M$ represents hours of work and $b >> 0_M$ is a positive vector that could perhaps represent the maximum hours that could be supplied of the various types of household labour services. This approach is pursued in Riddell [1983]. However, it will be difficult to come to an agreement on just what value we should take for the vector b (particularly if members of the household are holding multiple jobs). Hence the resulting Konüs Cost of Living index would be somewhat arbitrary, and should therefore not be used as an inflation measure.[28].

However, it is still possible to use index number techniques in order to obtain approximations to the change in the household's welfare which occurred going from period 0 to period 1. Essentially, what we shall do is measure welfare changes in terms of proportional changes in consumption goods.

First we define the household's conditional consumption deflation function D by

$$D(u,x,y) \equiv \max_k \left\{ k: F(x/k,y) \geq u, \; k > 0 \right\} \tag{86}$$

where $u > 0$ is a reference utility level, $x > 0_N$ is a consumption vector, $y \leq 0_M$ is a vector of labour supplies indexed negatively and F is the household utility function. If $D(u,x,y,)$ $> 1 \, (<1)$, then the household joint consumption labour supply vector yields a utility level greater (less) than u, while if $D(u,x,y) = 1$, then (x,y) yields precisely the utility level u. In general $D(u,x,y)$ tells us the proportion $k^* > 0$ that we have to deflate the consumption vector x so that $(x/k^*,y)$ will yield utility level u for the household.

A Malmquist [1953] consumption quantity index may now be defined in the usual manner:

$$Q(x^0,y^0,x^1,y^1,u) \equiv D(u,x^1,y^1)/D(u,x^0,y^0). \tag{87}$$

A geometric interpretation of the quantity index defined by (87) may be found on Figure 3 for the case of one consumption good (N = 1) and one type of labour supply (M = 1).

Figure 3   The Malmquist Quantity Index in the Labour Supply Context.

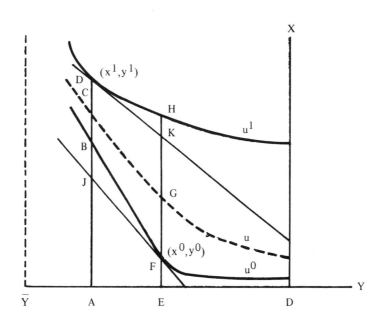

For the reference utility level u (which is between $u^0 \equiv F(x^0,y^0)$ and $u^1 \equiv F(x^1,y^1)$), it can be verified that $Q(x^0,y^0,x^1,y^1,u) = [DA/CA]/[FE/GE]$.

As usual, it is useful to let the references utility level u in (87) be either the base utility level $u^0 \equiv F(x^0,y^0)$ or the period 1 utility level $u^1 \equiv F(x^1,y^1)$. Thus define the *Laspeyres-Malmquist consumption quantity index* by $Q(x^0,y^0,x^1,y^1,F(x^0,y^0))$ and the *Paasche-Malmquist consumption quantity index* by $Q(x^0,y^0,x^1,y^1,F(x^1,y^1))$. In Figure 2, the first of these indexes is DA/BA while the second is HE/FE.

Many of the theorems that were stated in Section 3 go through in the present context. In particular, a counterpart to Theorem 7 is true, so if the reference utility level in (87) is between u^0 and u^1, $Q(x^0,y^0,x^1,y^1,u) > 1$ ($<1, = 1$) indicates that household utility has increased (decreased, remained constant) going from period 0 to period 1.

We also obtain the following counterpart to Theorem 9.

Theorem 21

Suppose F satisfies the modified Conditions I and the household's observed period 0 and period 1 choices (x^0, y^0) and (x^1, y^1) are consistent with utility maximizing or cost minimizing behaviour; i.e., x^0, y^0 satisfies

$$p^0 \cdot x^0 + w^0 \cdot y^0 = C[F(x^0, y^0), p^0, w^0] \tag{88}$$

where (p^0, w^0) are the observed positive period 0 prices and C is the cost function defined by (84), and (x^1, y^1) satisfies

$$p^1 \cdot x^1 + w^1 \cdot y^1 = C[F(x^1, y^1), p^1, w^1] \tag{89}$$

where (p^1, w^1) are the observed positive period 1 prices. Then provided that $p^0 \cdot x^0 + w^0 \cdot (y^0 - y^1) > 0$ and $p^1 \cdot x^1 + w^1 \cdot (y^1 - y^0) > 0$,

$$Q(x^0, y^0, x^1, y^1, F(x^0, y^0)) \leq p^0 \cdot x^1 / [p^0 \cdot x^0 + w^0 \cdot (y^0 - y^1)] \equiv \alpha \text{ and} \tag{90}$$

$$Q(x^0, y^0, x^1, y^1, F(x^1, y^1)) \geq [p^1 \cdot x^1 + w^1 \cdot (y^1 - y^0)] / p^1 \cdot x^0 \equiv \beta. \tag{91}$$

The bounds in (90) and (91) may be illustrated by referring to Figure 3. (90) becomes DA/BA $\leq$ DA/JA and (91) becomes HE/FE $\geq$ KE/FE. As the reader can observe, the bounds are rather close to the appropriate theoretical index. It is also true that the bounds α and β will often be close to each other. Hence the following counterpart to Theorem 10 is of some practical interest.

Theorem 22

Assume the regularity conditions of the previous theorem and define the base utility $u^0 \equiv F(x^0, y^0)$ and the period 1 utility level $u^1 \equiv F(x^1, y^1)$. Then there exists a reference utility level u^* between u^0 and u^1 such that the Malmquist consumption quantity index $Q(x^0, y^0, x^1, y^1, u^*)$ lies between α and β, the bounds defined in (90) and (91).

Thus if α and β are close, we will be able to obtain a good estimate of $Q(x^0, y^0, x^1, y^1, u^*)$ by averaging α and β.

For an empirical implementation of the above material, see Riddell [1983].

12. Durables in the Cost-of-Living Index

The treatment of consumer durables in the CPI and in the true cost-of-living index is an interesting and controversial issue.[29]

The basic issues can readily be explained. Consider a durable good that can be purchased at the beginning of the current period at the spot price q^0. A consumer can purchase this good at the beginning of the period, use the good during period 0, but because of the good's durable nature, some of it will be left over at the beginning of the following period. The consumer could sell his used durable good at a (possibly hypothetical) second-hand market at an **expected** price of q^{01}. Assuming that the consumer can lend or borrow at the rate of return r, we may follow Hicks [1946] and compute the present value of the cost of buying one unit of the good, using it for one period, and selling it next period. The resulting **user cost** p is

$$p = q^0 - q^{01}/(1+r) = [q^0 r + (q^0 - q^{01})]/[1+r]. \qquad (92)$$

The first term on the right-hand side of (82) is an interest cost while the second term combines the effects of anticipated capital gains and depreciation. We can separate out these two effects if we let q^1 be the spot price of a new unit of the durable that the consumer expects to prevail during period 1. Hence we may write $(q^0 - q^{01}) = (q^1 - q^{01}) - (q^1 - q^0)$ = depreciation - capital gains. Thus the user cost (92) becomes

$$p = (1+r)^{-1}[q^0 r + (q^1 - q^{01}) - (q^1 - q^0)]. \qquad (93)$$

The above derivation of the user cost of a durable in discrete time was essentially obtained by Diewert [1974; p.504] and Pollak [1975b]. The term $(1+r)^{-1}$ may strike the reader as being a bit odd, but we need it so that when $q^{01} = 0$ (so that the good is actually a non-durable), then the user cost p collapses to the period 0 purchase price q^0.

Unfortunately, there are many problems associated with the user cost formulae (92) or (93).

The first problem is that the prices q^{01} and q^1 are not market prices - they are the consumer's period 0 expectations of what the market prices will actually be in period 1. It is not realistic to assume that our non-renting consumer will be able to accurately forecast these future prices. If consumers were able to accurately forecast future prices, then the existing rental market price for the services of the durable would equal the user cost p defined by (92) where the expected price q^{01} in (92) is replaced by the observed *ex-post* market price. Hendershott [1980; 406] demonstrates that rental prices for houses do not track ex post user costs for housing very closely.[30] However, all is not lost if there are some rental markets for the class of durable goods under consideration.

If we do have some rental market information on the durable then we could assume that the rental price equals the user cost p that appears in (92) and we could use (92) to **solve** for the expected price q^{01} in terms of the observed market prices p, q^0 and r. Thus expectational information gleamed in this way for classes of durables that have rental markets could be applied to forecast expectations of future prices for classes of "similar" durables that do not have rental markets.

Most goods can only be purchased in integral numbers, and for most goods, this does not cause major problems. However, some durable goods such as cars and houses may be purchased only in integer units, and such purchases would form a large share of the consumer's total expenditure. Hence we cannot neglect the lumpiness problem for such classes of durables. How may we apply traditional "continued" utility and index number theory to this situation?

A possible solution is illustrated in Figure 4 below. Assume x_2 represents units of a "continuous" good, ice cream say, while x_2 represents the number of television sets that a household holds during a period.

In Figure 4, we have graphed two indifference "curves" for the household. Only the "kinky" points on the curves plus the line segments parallel to the x_2 axis are actual feasible choices for the household, but for all practical purposes, we can replace the original preferences defined only over integer combinations of TV sets by continuous preferences with "kinks".[31] The resulting preference function F(x) may be treated in the normal manner as far as index number theory is concerned. Note that the economic effect of the "kinks" will be to make the consumer change his durable holdings only after relatively large changes in the rental prices of the durables relative to non-durable goods; i.e., responses will be "sticky". This point should be taken into account in econometric work, but it need not concern us from the viewpoint of index number theory.

Another difficulty (which **does** create problems for us from the viewpoint of index number theory) is that the expected buying and selling price of the durable may not be the same; i.e., there may be significant transactions costs associated with buying and selling units of the durable. The effect of this difference in buying and selling prices for the durable will be to put a "link" in the consumer's budget constraint around his initial holdings of the durable. See Figure 5 below.

Figure 4.

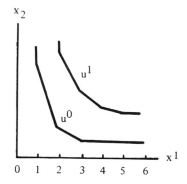

Figure 5.

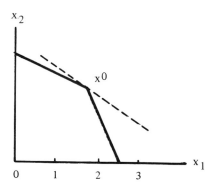

If the consumer is observed at x^0 during period 0, the correct price of the durable to use in an index number comparison lies somewhere between the buying and selling price. For our purposes, we shall probably have to settle for taking an average of the two prices. Note that the effect of this difference in buying and selling prices will again have the effect of making the consumer's demand to hold durables "sticky" around his initial holdings.[32]

Differences in borrowing and lending rates for a household may also have the effect of introducing a "kink" into the budget set when the purchase of a large consumer durable is contemplated. Deaton and Muellbauer [1980; Ch.13] discuss the effect of a borrowing constraint, and in fact, they have an excellent discussion of the problems involved in modelling the demand for consumer durables.

The final problem that we wish to discuss is the problem of calculating the user cost of a durable when there are tax considerations involved.

At the marginal investment point, we assume that the durable holder invests in a market asset that earns a before tax rate of return r. Suppose that consumer's marginal tax rate is τ. The present value of the cost of using the durable for one period is

$$p \equiv q^0 - \frac{q^{01}}{1+(1-\tau)r} + \frac{\tau \alpha(q^{01}-q^0)}{1+(1-\tau)r} - \frac{\tau \beta r_M q^0}{1+(1-\tau)r} + \frac{tq^0}{1+(1-\tau)r} - \frac{\tau \delta q^0}{1+(1-\tau)r} \qquad (94)$$

where q^0 and q^{01} are the same as before, α is the proportion of capital gains on the durable that is taxable, β is the proportion of (mortgage) interest that is deductible from taxable income (and r_M is the appropriate mortgage interest rate), t is a user tax rate on holdings of the durable (e.g., a property tax), δ is the depreciation rate that is allowed for taxation purposes, and all taxes are assumed to be payable in the following period (and hence they are discounted). If we assume $\alpha = \beta = \delta = 0$, then (94) reduces to

$$p = [(1-\tau)rq^0 + tq^0 - (q^{01}-q^0)]/[1+(1-\tau)r]. \qquad (95)$$

Thus the higher is the marginal tax rate , the lower will be the user cost p. If the durable holder is a borrower (at the mortgage rate r_M say), then the discount factor in (94) and (95) must be replaced by $1 + r_M$. Under these conditions, (95) becomes

$$p = [r_M q^0 + t q^0 - (q^{01} - q^0)]/[1 + r_M]. \qquad (96)$$

which will be much higher than the user cost defined by (95).

Thus the tax and financial situation of the consumer plays an essential role in the calculation of user costs for durables. This is a very unfortunate result, since the informational requirements for implementing a user cost formula such as (94) are very high. In particular, the expected price q^{01} that appears in (94) is not directly observable. Also it may be very difficult to determine precisely what is the appropriate opportunity cost of the marginal investment (r) for the consumer under consideration. These problems are particularly acute in the case of housing, since housing expenditures are generally a large proportion of a typical household budget. In order to avoid the difficult measurement problems associated with the user cost approach, Gillingham [1982] suggests essentially that the price quantity data that pertains to the rental segment of the housing market be extrapolated to the entire housing market. The problem with this rather sensible suggestion is that the rental segment of the housing market is generally not representative of the entire housing market. A possible reason for this non-representativeness emerges if we compare the user cost for a house for a rich individual with a high marginal tax rate δ (see (95)) with the user cost formula for a less well-off individual who has no non-labour income and holds a mortgage on his house (recall (96)): the same house will cost the rich individual far less in terms of user cost than the poor individual. Furthermore, the tax laws in most Western countries will generally make it more profitable for a rich individual to own and live in his house rather than rent out his house and live in a rental house of comparable quality. Thus the rental price information for the rental portion of the housing market will not be representative for the market as a whole (due to these tax considerations). Moreover, as nominal interest rates and marginal tax rates changed over time, we could expect the price of rental housing to systematically deviate from the appropriate user cost index of non-rental housing. Thus I do not believe that the informational difficulties that are inherent in the user cost formula (94) can be avoided. For further discussion on the role of tax considerations in the construction of user costs, see Darrough [1983].

We conclude this section with a reminder that all of our bounds on the consumer's true cost-of-living index rested on the assumption of utility maximizing behaviour subject to a budget constraint. The prices that appear in the budget constraint are observable market prices in the case of non-durable goods, but for a durable good which is owned by the consumer, the appropriate price must be an *ex ante* user cost of the form (94), which depends on the (unobservable) anticipated price of the depreciated durable which is expected to prevail in the following period. It is not in general appropriate to use an *ex post* user cost formula of the form (94) where the anticipated price q^{01} is replaced by an (observable) market price, since the resulting *ex post* user cost may well be negative if there is an unanticipated inflation in the price of the durable. Thus the existence of unanticipated inflation (or deflation) and the non-neutrality of the income tax with respect to the treatment of durables make it very difficult to construct a Konüs cost-of-living index (or bounds to it) for a household. Thus there are costs due to unanticipated inflation and the non-neutrality of the present tax system.

13. The New Goods Problem

The standard approach to the new goods problem in the context of consumer theory dates back to Hicks [1940]: the period after a new good appears (period 1 say), we attempt to impute a price for the new good in period 0 which would just make the consumer's demand for the good equal to 0 in period 0. The details of an econometrically implementible approach are outlined in Diewert [1980; pp.501-503].

However, a simpler method that is more practical is readily available: as soon as a new good appears, start collecting price and quantity information on it. Since initially the quantity purchased will be low and the price will often be rather high, a quick introduction of the good into the CPI universe of prices would solve most of the practical problems associated with the current neglect of the quality change problem in official CPIs.[33] It is true that many new goods quickly disappear, but this causes no particular theoretical problems. However, we must concede that linking the prices of similar new goods that vary in quality poses some practical problems.

14. Conclusion

The main conclusions which emerge from this paper are listed below.

(i) In addition to the usual Laspeyres based CPI, a Paasche based CPI should also be published as frequently as possible, since an appropriate true cost-of-living index lies between a Paasche and Laspeyres index. This means that household surveys where quantity information is collected should be undertaken more frequently, say every second year.

(ii) CPIs should be constructed on a more disaggregated basis (by household demographic characteristics, by income and by region).

(iii) Labour supply and leisure should be introduced into the CPI framework on an experimental basis. The consumer-worker's income tax position will play an important role here.

(iv) Intertemporal CPIs are too problematical to be introduced at this time.

(v) The treatment of consumer durables, particularly housing, is not very satisfactory at present. Various user cost and rental equivalent alternatives should be tried on an experimental basis. The importance of the household's tax and financial position should not be overlooked.

(vi) The treatment of seasonal commodities is also unsatisfactory. A more satisfactory treatment of seasonal commodities from the viewpoint of economic theory is outlined in Diewert [1983b].

(vii) New goods should be introduced into the CPI universe of prices as soon as possible. The neglect of new goods provides an upward bias to the existing CPI of a possibly major magnitude.

If the above recommendations are implemented, then not only will we have much more accurate information on how inflation affects different consumer groups, we will also be able to simulate how changes in tax policy affect the welfare of the different household groups.

Appendix: Proofs of Theorems

Proof of Theorem 1: Define $C(u,p) \equiv uP(p^*, p,u)$ for $u > 0$ and $p >> 0_N$. Let $u > 0$, $p^0 >> 0_N$ and $p^1 >> 0_N$. Then

$$C(u,p^1)/C(u,p^0) \equiv uP(p^*,p^1,u)/uP(p^*,p^0,u)$$

$$= P(p^*,p^1,u)/P(p^*,p^0,u)$$

$$= P(p^0,p^*,u)P(p^*,p^1,u)$$

using the time reversal property (ii)

$$= P(p^0,p^1,u)$$

using the circularity property (iii).

Note that properties (ii) and (iii) imply $P(p^*,p^*,u) = 1$. Hence $C(u,p^*) \equiv uP(p^*,p^*,u) = u$ for all $u > 0$ which is (3).

The proof of the converse part of the theorem is straightforward. I owe this method of proof to David Donaldson.

Proof of Theorem 14: Define $h(\lambda) \equiv P_D(p^0,p^1, (1-\lambda)u^0 + \lambda u^1)$ for $0 \leq \lambda \leq 1$. Note that $h(0) = P_D(p^0,p^1,u^0)$ and $h(1) = P_D(p^0,p^1,u^1)$. There are 24 possible *a priori* inequality relations that are possible between the four numbers $h(0)$, $h(1)$, $\bar{P}_L$ and $\bar{P}_P$. However, (51) and (52) imply that $h(0) \leq \bar{P}_L$ and $\bar{P}_P \leq h(1)$. This means that there are only six possible inequalities between the four numbers:

(1) $h(0) \leq \bar{P}_L \leq \bar{P}_P \leq h(1)$, (2) $h(0) \leq \bar{P}_P \leq \bar{P}_L \leq h(1)$, (3) $h(0) \leq \bar{P}_P \leq h(1) \leq \bar{P}_L$, (4) $\bar{P}_P \leq h(0) \leq \bar{P}_L \leq h(1)$, (5) $\bar{P}_P \leq h(1) \leq h(0) \leq \bar{P}_L$ and (6) $\bar{P}_P \leq h(0) \leq h(1)$ $\leq \bar{P}_L$. Since the individual cost functions $C^h(u_h, p^t)$ are continuous in u_h, it can be seen that $P_D(p^0, p^1, u)$ defined by (49) is continuous in the vector of utility variables u. Hence $h(\lambda)$ is a continuous function for $0 \leq \lambda \leq 1$ and assumes all intermediate values between $h(0)$ and $h(1)$. By inspecting cases (1) to (6) above, it can be seen that we can choose λ between 0 and 1 (call this number λ^*) so that $\bar{P}_L \leq h(\lambda^*) \leq \bar{P}_P$ for case (1) or so that $\bar{P}_P \leq h(\lambda^*) \leq \bar{P}_L$ for cases (2) to (6). Now define $u^* \equiv (1-\lambda^*)u^0 + \lambda^* u^1$ and the proof is complete.

Proof of Theorem 15: Using expression (54) for $P_{PP}(p^0, p^1, u^0)$ and $P_{PP}(p^0, p^1, u^1)$, noting that the shares $s^h(u^0, p^0)$ and $s^h(u^1, p^0)$ defined by (55) are non-negative and sum to 1, and using Theorem 2, we may readily establish the left inequality in (57) and the right inequality in (58).

Consider now the right inequality in (57). This follows readily from definition (53), the equalities $C^h(u_h^0, p^0) = p^0 \cdot x_h^0$ for $h = 1, \ldots, H$ and the inequalities

$$C^h(u_h^0, p^1) \equiv \min_x \left\{ p^1 \cdot x : F^h(x) \geq F^h(x_h^0) \equiv u_h^0 \right\}$$

$$\leq p^1 \cdot x_h^0$$

(A1)

which follow since x_h^0 is feasible for the minimization problem.

The left inequality in (58) follows from definition (53) when $u = u^1$, the equalities $C^h(u_h^1, p^1) = p^1 \cdot x_h^1$, and the inequalities

$$C^h(u_h^1, p^0) \equiv \min_x \left\{ p^0 \cdot x : F^h(x) \geq F^h(x_h^1) \equiv u_h^1 \right\}$$

$$\leq p^0 \cdot x_h^1.$$

(A2)

Proof of Theorem 16: The proof of this theorem is identical to the proof of Theorem 14 if we replace P_D by P_{PP}, $\bar{P}_L$ by P_L and $\bar{P}_P$ by P_P.

Proof of Theorem 17: First note that for each household h,

$$D^h(u_h^0, x_h^1) \equiv \max_k \{k: F^h(x_h^1/k) \geq u_h^0\}$$

$$= k_h^1 \text{ where } F^h(x_h^1/k_h^1) = u_h^0.$$

Thus

$$p^0 \cdot x_h^0 = C^h(u_h^0, p^0) \tag{A3}$$

$$= \min_x \{p^0 \cdot x: F^h(x) \geq u_h^0\}$$

$$\leq p^0 \cdot x_h^1/k_h^1$$

since x_h^1/k_h^1 is feasible for the cost minimization problem. Thus (A3) yields the following inequality:

$$D(u_h^0, x_h^1) \leq p^0 \cdot x_h^1/p^0 \cdot x_h^0 \qquad \text{for } h = 1,\dots,H.$$

Repeating the above argument interchanging the superscripts 0 and 1 yields the following inequality:

$$D(u_h^1, x_h^0) \leq p^1 \cdot x_h^0/p^1 \cdot x_h^1 \qquad \text{for } h = 1,\dots,H. \tag{A5}$$

From definition (60), we have

$$W(x^0, x^1, u^0) \equiv \Sigma_{h=1}^{H} \beta_h^0 D^h(u_h^0, x_h^1) / D^h(u_h^0, x_h^0)$$

$$= \Sigma_{h=1}^{H} \beta_h^0 \ D^h(u_h^0, x_h^1) \qquad \text{since } D^h(u_h^0, x_h^0) = 1$$

$$\leq \Sigma_{h=1}^{H} \beta_h^0 \ p^0 \cdot x_h^1 / p^0 \cdot x_h^0$$

using (A4) and $\beta_h^0 \geq 0$, which establishes (63). Similarly

$$W(x^0, x^1, u^0) \equiv \Sigma_{h=1}^{H} \beta_h^1 D^h(u_h^1, x_h^1) / D^h(u_h^1, x_h^0)$$

$$= \Sigma_{h=1}^{H} \beta_h^1 \ 1 / D^h(u_h^1, x_h^0) \qquad \text{since } D^h(u_h^1, x_h^1) = 1$$

$$\geq \Sigma_{h=1}^{H} \beta_h^1 \ p^1 \cdot x_h^1 / p^1 \cdot x_h^0 \qquad \text{using (A5) and } \beta_h^1 \geq 0.$$

Proof of Theorem 18: Define $h(\lambda) \equiv W(x^0, x^1, (1-\lambda)u^0 + \lambda u^1)$ and repeat the proof of Theorem 14, where $\Sigma_{h=1}^{H} \beta_h^0 Q_L^h$ replaces $\bar{P}_L$ and $\Sigma_{h=1}^{H} \beta_h^1 Q_P^h$ replaces $\bar{P}_P$.

Proof of Theorem 19: First note that

$$C(F(x^0), p^0) \equiv \min_x \left\{ p^0 \cdot x : F(x) \geq F(x^0) \right\}$$

$$= p^0 \cdot x^0 \qquad \text{by (5)}$$

$$= \Sigma_{m=1}^{M} p_m^0 \cdot x_m^0$$

$$= \min_{x_1} \left\{ p_1^0 \cdot x_1 + \sum_{m=2}^{M} p_m^0 \cdot x_m^0 : F(x_1, x_2^0, ..., x_M^0) \right.$$

$$\left. \geq F(x_1^0, x_2^0, ..., x_M^0) \right\}$$

$$\equiv C^1(F(x^0), p_1^0, x^0) + \Sigma_{m=2}^{M} p_m^0 \cdot x_m^0 .$$

Hence $C^1(F(x^0), p_1^0, x^0) = p_1^0 \cdot x_1^0$. In a similar manner, we obtain the following equalities:

$$C^m(F(x^0), p_m^0, x^0) = p_m^0 \cdot x_m^0, \qquad m = 1,2,\ldots,M \text{ and} \qquad \text{(A6)}$$

$$C^m(F(x^1), p_m^1, x^1) = p_m^1 \cdot x_m^1, \qquad m = 1,2,\ldots,M. \qquad \text{(A7)}$$

Since x_m^0 is feasible for the conditional cost minimization problem defined by $C^m(F(x^0), p_m^1, x^0)$, we have

$$C^m(F(x^0), p_m^1, x^0) \leq p_m^1 \cdot x_m^0, \qquad m = 1,\ldots,M. \qquad \text{(A8)}$$

Since x_m^1 is feasible for $C^m(F(x^1), p_m^0, x^1)$,

$$C^m(F(x^1), p_m^0, x^1) \leq p_m^0 \cdot x_m^1, \qquad m = 1,\ldots,M. \qquad \text{(A9)}$$

The definition of $P^m(p_m^0, p^1, u^0, x^0) \equiv C^m(F(x^0), p_m^1, x^0)/C^m(F(x^0), p_m^0, x^0)$, (A6) and (A8) yield (67) while $P^m(p_m^0, p_m^1, u^1, x^1) \equiv C^m(F(x^1), p_m^1, x^1)/C^m(F(x^1), p_m^0, x^1)$, (A7) and (A9) yield (68).

Footnotes

[1] The conceptual framework for the Canadian CPI is nicely explained in Statistics Canada [1982]. The CPI and the Implicit Consumption Price Index for Canada are compared and contrasted in Loyns [1972], who was mainly interested in their inflation measuring capabilities. My focus will be more welfare-oriented.

[2] Unfortunately, much of the material presented in this paper is a bit technical. Two useful references that lead the reader into the technical aspects of index number and growth measurement theory in a gentle fashion are Allen [1975], and Usher [1980]. More technical discussions may be found in Konüs [1924], Samuelson [1947; pp.146-162], Malmquist [1953], Pollak [1971], Afriat [1977] and Diewert [1981].

[3] Notation: x^T denotes the transpose of the column vector x, $p^Tx = p \cdot x \equiv \Sigma_{n=1}^{N} p_n x_n$ denotes the inner product of the vectors p and x, $x \geq 0_N$ means each component of the vector x is non-negative, $x >> 0_N$ means each component if positive, and $x > 0_N$ means $x \geq 0_N$ but $x \neq 0_N$.

[4] F is a function defined over the non-negative orthant $\{x: x \geq 0_N\}$ that has the following properties: (i) continuity, (ii) increasingness; i.e., if $x'' >> x' \geq 0_N$, then $F(x'') > F(x')$, (iii) quasiconcavity; i.e., for each utility level u, the upper level set $L(u) \equiv \{x: F(x) \geq u\}$ is convex, (iv) $F(0_N) = 0$ and (v) F(x) tends to $+\infty$ as the components of x all tend to $+\infty$.

[5] C(u,p) is defined for $u \geq 0$, $p >> 0_N$ and has the following properties: (i) it is continuous, (ii) C(0,p) = 0 for every $p >> 0_N$, (iii) for every $p >> 0_N$, C(u,p) is increasing in u and C(u,p) tends to $+\infty$ as u tends to $+\infty$, (iv) C(u,p) is positively linearly homogeneous in p for fixed u, i.e., for $u \geq 0$, $p >> 0_N$, $\lambda > 0$, $C(u,\lambda p) = \lambda C(u,p)$, (v) C(u,p) is concave in p for fixed u, (vi) C(u,p) is increasing in p for fixed $u > 0$, i.e., if $p'' >> p' >> 0_N$, $u > 0$, then $C(u,p'') > C(u,p')$ and (vii) C is such that the function $F^*(x) \equiv \max_u \{u: p \cdot x \geq C(u,p)$ for every $p >> 0_N$, $u \geq 0\}$ is continuous for $x \geq 0_N$.

[6] This is a version of the Shephard [1953] Duality Theorem; see Diewert, [1982].

[7] The term is due to Samuelson [1974].

[8] Throughout this section, we assume that F satisfies Conditions I. Many of the theorems in this section can be proven under much weaker regularity conditions; e.g., see Diewert [1981].

[9] See also Diewert [1973] and Varian [1982].

[10] If $u^0 = u^1$, then $u^* = u^0 = u^1$.

[11] Thus c^r is a flexible functional form to use Diewert's [1974] terminology.

[12] This corresponds to the terminology used in Christensen, Cummings and Jorgenson [1980]. Diewert [1981; p.187] called P_0 the Törnqvist price index, but the term translog price index seems to be more descriptive.

[13] The Allen quantity index is closely related to: (i) Samuelson's [1974] money metric scaling for a consumer's utility function, and (ii) Hicks' [1941-42] **consumer surplus** measures, which are defined in terms of differences of cost functions rather than ratios of cost functions.

[14] The appropriate regularity conditions are listed in Diewert [1982; p.560], and references to the literature on duality theorems between F and D may be found there also. Essentially, D(u,x) has the same regularity properties as the cost function C(u,p) where x replaces p, except that D(u,x) decreases in u while C(u,p) increases in u.

[15] Theorem 7 is not necessarily true if $u > \max\left\{u^0, u^1\right\}$ or if $u < \min\left\{u^0, u^1\right\}$.

[16] Related approximation theorems have been obtained by Samuelson and Swamy [1974] and Vartia [1978]. It should be noted that Diewert's [1978] results were derived using some results due to Vartia [1976].

[17] For example, see Diewert [1978; p.894], Généreux [1983] and Szulc [1983].

[18] In fact Allen and Diewert [1981; p.435] provide an even stronger case for the use of the Fisher ideal formula, since it is the only superlative index number formula that is consistent with both the Hicks [1946; pp. 312-313] and Leontief [1936; pp. 54-57] composite commodity theorems.

[19] We cannot even evaluate how good or bad the approximation is unless we are also given current period quantity information x^t.

[20] Pollak [1975a] demonstrates that it is difficult if not impossible to combine the subindexes into the true cost-of-living index under general conditions on the underlying preferences. This is not exactly the relevant issue, since we cannot calculate the true cost-of-living index anyway in general. However, if we can use the subindexes to form a close approximation to the overall true cost-of-living index, then this is all that we require. Of course, under restrictive assumptions on preferences, subindexes can be combined to give precisely the correct overall cost-of-living index. The first result of this type was obtained by Shephard [1953], who assumed a special structure of preferences that is now called homothetic separability. For generalizations of the Shephard result and reference to the literature, see Blackorby, Primont and Russell [1978; ch. 9].

[21] In Pollak [1975; p.145], C^m is called a generalized conditional expenditure function for category m.

[22] $C^m(u, p_m, x)$ is non-decreasing in u and non-increasing in the components of x. See McFadden [1978] for a detailed analysis of the properties of C^m.

[23] Remember that $C^m(u, p_m, x)$ does not actually depend on the mth subvector in x, x_m, and hence $P^m(p_m^0, p_m^1, u^*, x^*)$ does not actually depend on x_m^*, the mth subvector in $x^* \equiv (x_1^*, x_2^*, ..., x_M^*)$.

[24] Vartia [1976] defines an index number formula to be **consistent in aggregation** if the value of the index calculated in two stages coincides with the value of the index calculated in a single stage. A careful analysis of this concept may be found in Blackorby and Primont [1980]. Empirical evidence on the closeness of two-stage aggregates with the corresponding single-stage aggregates may be found in Diewert [1978, 1983b].

[25] If we are willing to use econometric techniques, then it is not necessary to assume intertemporal additivity in order to estimate the consumer's intertemporal preference function; i.e., see Diewert [1974] and Darrough [1977]. This suggests that it may be possible to adapt the usual index number techniques to the intertemporal context as well.

[26] In the consumer context, see Ruggles [1967], and in the producer context, see Denny and Fuss [1981] and Caves, Christensen and Diewert [1982a].

[27] We are assuming that the household derives disutility from supplying additional hours of work. This may not be true for (x,y) vectors where y is close to 0_M. Formally, we assume that F satisfies Conditions I except that now the domain of definition of F is (x,y): $x \geq 0_N$, $\bar{y} \leq y \leq 0_M$ where $\bar{y} \leq 0_M$.

[28] However, it could still be used as a deflator for the household's nominal "full" income ratio, $(p^1 \cdot x^1 + w^1 \cdot b)/(p^0 \cdot x^0 + w^0 \cdot b)$, to form Pollak implicit quantity indexes as was done in the beginning of Section 3 above. On the concept of "full" income, see Becker [1965].

[29] Recent papers on this issue include McFadyen and Hobart [1978], Rymes [1979], Blinder [1980], Hendershott [1980], Hughes [1980], Gordon [1981], Dougherty and Van Order [1982] and Gillingham [1982].

[30] In fact when one works with user cost formulae of the type defined by (93) and evaluates the expected prices by using *ex post* market prices, one will often find negative user costs for housing.

[31] In technical terms, we replace the original preferences by the convex free disposal hull of the original preferences.

[32] Other types of transactions costs will also have this effect.

[33] Comprehensive accounts of the quality change problem may be found in Triplett [1982] and Hodgins [1982].

[*] This research was supported by Statistics Canada and the SSHRC of Canada. Neither institution is responsible for the views expressed here. The author is indebted to B. Balk, M. Darrough, David Donaldson, W. Eichhorn, R. Gillingham, B.J. Szulc and J. Weymark for helpful comments.

References

Afriat, S.N. [1967], "The Construction of Utility Functions from Expenditure Data", *International Economic Review* 8, pp.67-77.

_____ [1972], "The Theory of International Comparisons of Real Income and Prices", pp.13-69 in D.J. Daly (ed.), *International Comparisons of Prices and Output*, New York: National Bureau of Economic Research.

_____ [1977], *The Price Index*, London: Cambridge University Press.

Allen, R.G.D. [1949], "The Economic Theory of Index Numbers", *Economica* N.S. 16, pp.197-203.

_____ [1975], "Index Numbers in Theory and Practice", London: Macmillan.

Allen, R.C. and W.E. Diewert [1981], "Direct versus Implicit Superlative Index Number Formulae", *The Review of Economics and Statistics* 63, pp.430-435.

Becker, G.S. [1965], "A Theory of the Allocation of Time", *The Economics Journal*, 75, pp.493-517.

Blackorby, C. and D. Donaldson [1983], "Preference Diversity and Aggregate Economic Cost-of-Living Indexes", *Price Level Measurement: Proceedings From a Conference Sponsored by Statistics Canada.*

_____ D. Primont and R.R. Russell [1978], *Duality Separability and Functional Structure: Theory and Economic Applications*, New York: North-Holland.

_____ and D. Primont [1980], "Index Numbers and Consistency in Aggregation", *Journal of Economic Theory*, 22, pp.87-98.

Blinder, A.S. [1980], "The Consumer Price Index and the Measurement of Recent Inflation, *Brookings Papers on Economic Activity*, pp.539-565.

Caves, D.W., L.R. Christensen and W.E. Diewert [1982a], "Multilateral Comparisons of Output, Input and Productivity using Superlative Index Numbers", *The Economic Journal* 92, pp.73-86.

_____ [1982b], "The Economic Theory of Index Numbers and the Measurement of Input, Output and Productivity", *Econometrica* 50, pp.1393-1414.

Christensen, L.R., D. Cummings and D.W. Jorgenson [1980], "Economic Growth, 1947-73: An International Comparison", pp.595-691 in *New Developments in Productivity Measurement and Analysism* J.W. Kendrick and B.N. Vaccara (eds.), National Bureau of Economic Research, Studies in Income and Wealth, Vol. 44, Chicago: University of Chicago Press.

Darrough, N.M. [1977],"A Model of Consumption and Leisure in an Intertemporal Framework: A Systematic Treatment using Japanese Data", *International Economic Review*, 18, 677-696.

Darrough, M. [1983], "The Treatment of Housing in a Cost of Living Index: Rental Equivalence and User Cost", *Price Level Measurement: Proceedings From a Conference Sponsored by Statistics Canada.*

Deaton, Angus and J. Muellbauer [1980], *Economics and Consumer Behavior*, London: Cambridge University Press.

Denny, M. [1974], "The Relationship between Functional Forms for the Production System", *Canadian Journal of Economics* 7, pp.21-31.

_____ and M. Fuss [1981], "Intertemporal Changes in Regional Productivity in Canadian Manufacturing," *The Canadian Journal of Economics* 14, pp.390-408.

_____ [1983], "Regional Price Indexes: The Canadian Practice and Some Potential Extensions," *Price Level Measurement: Proceedings From a Conference Sponsored by Statistics Canada.*

Diewert, W.E. [1973], "Afriat and Revealed Preference Theory", *Review of Economic Studies* 40, pp.419-425.

_____ [1974], "Intertemporal Consumer Theory and the Demand for Durables", *Econometrica* 42, pp.497-516.

_____ [1976], "Exact and Superlative Index Numbers", *Journal of Econometrics* 4, pp.115-145.

_____ [1978], "Superlative Index Numbers and Consistency in Aggregation", *Econometrica* 46, pp.883-900.

_____ [1980], "Aggregation Problems in the Measurement of Capital", pp.433-528 in *The Measurement of Capital*, D. Usher (ed.), Chicago: The University of Chicago Press.

_____ [1981], "The Economic Theory of Index Numbers: A Survey," pp.163-208 in *Essays in the Theory and Measurement of Consumer Behavior in Honour of Sir Richard Stone*, Angus Deaton, (ed.), Cambridge, England: Cambridge University Press.

_____ [1982a], "Duality Approaches to Microeconomic Theory", pp.535-599 in *Handbook of Mathematical Economics*, Vol. II, K.J. Arrow and M.D. Intriligator, (eds.), Amsterdam: North-Holland.

_____ [1983b], "The Treatment of Seasonality in a Cost-of-Living Index", *Price Level Measurement: Proceedings From a Conference Sponsored by Statistics Canada.*

_____ [1983], "The Theory of the Output Price Index and the Measurement of Real Output Change", *Price Level Measurement: Proceedings From a Conference Sponsored by Statistics Canada.*

Dougherty, Ann and R. Van Order [1982], "Inflation, Housing Costs and the Consumer Price Index", *American Economic Review* 72, pp.154-164.

Eichhorn, W. [1976], "Fisher's Tests Revisited", *Econometrica* 44, pp.247-256.

_____ [1978], *Functional Equations in Economics*, Reading, Mass.: Addison-Wesley.

_____ and J. Voeller [1976], *Theory of the Price Index: Fisher's Test Approach and Generalizations*, Lecture Notes in Economics and Mathematical Systems, Berlin: Springer-Verlag.

_____ [1983], "The Axiomatic Foundations of Price Indexes and Purchasing Power Parities", *Price Level Measurement: Proceedings From a Conference Sponsored by Statistics Canada.*

Fisher, Irving [1922], *The Making of Index Numbers*, Boston: Houghton-Mifflin.

Généreux, P. [1983], "Impact of Alternative Formulae on the CPI", *Price Level Measurement: Proceedings From a Conference Sponsored by Statistics Canada.*

Gillingham, G. [1982], "Measuring the Cost of Shelter for Homeowners: Theoretical and Empirical Considerations", Division of Price and Index Number Research, Office of Prices and Living Conditions, BLS, Washington, D.C.

Gordon, R.J. [1981], "The Consumer Price Index: Measuring Inflation and Causing It", *The Public Interest* 63, pp.112-134.

Hendershott, P. [1980], "Real User Costs and the Demand for Single-Family Housing", *Brookings Papers on Economic Activity*, pp.401-444.

Hicks, J.R. [1940], "The Valuation of the Social Income", *Economica* 7, pp.105-124.

_____ [1941-42], "Consumers' Surplus and Index Numbers", *Review of Economic Studies*, 9, pp.126-137.

_____ [1946], *Value and Capital*, second edition, Oxford: Clarendon Press.

_____ [1958], "The Measurement of Real Income", *Oxford Economic Papers*, 10, pp. 125-162.

_____ [1961], "Measurement of Capital in Relation to the Measurement of other Economic Aggregates", in F.A. Lutz and D.C. Hague (eds.), *The Theory of Capital, London*: Macmillan.

_____ [1981], *Wealth and Welfare*, Cambridge, Mass.: Harvard University Press.

Hodgins, C.D. [1982], "Inflation Adjustments and Indexation: Is the Consumer Price Index an Appropriate Index?", mimeo, September 30.

Hughes, G.A. [1980], "Housing and the Tax System", pp.67-105, in *Public Policy and the Tax System*, G.A. Hughes, and G.M. Heal (eds.), London: George Allen and Unwin.

Jorgenson, D.W. and D.T. Slesnick [1983], "Individual and Social Cost-of-Living Indexes", *Price Level Measurement: Proceedings From a Conference Sponsored by Statistics Canada.*

Joseph, M.F.W. [1935-36], "Mr. Lerner's Supplementary Limits for Price Index Numbers", *Review of Economic Studies*, 3, pp.155-157.

Konüs, A.A. [1924], "The Problem of the True Index of the Cost of Living," translated in *Econometrica* 7, 1939, pp.10-29.

Leontief, W. [1936], "Composite Commodities and the Problem of Index Numbers", *Econometrica* 4, pp.39-59.

Lerner, A.P. [1935-36], "A Note on the Theory of Price Index Numbers", *Review of Economic Studies* 3, pp.50-56.

Loyns, R.M.A. [1972], *An Examination of the Consumer Price Index and Implicit Price Index as Measures of Recent Price Changes in the Canadian Economy*, Ottawa: Information Canada.

Malmquist, S. [1953], "Index Numbers and Indifference Surface", *Trabajos de Estadistica* 4, pp.209-242.

McFadden, D. [1978], "Cost, Revenue and Profit Functions", pp.3-109 in *Production Economics; A Dual Approach to Theory and Applications*, Vol. 1, M. Fuss and D. McFadden (eds.), Amsterdam: North-Holland.

McFadyen, S. and R. Hobart [1978], "An Alternative Measurement of Housing Costs and the Consumer Price Index", *Canadian Journal of Economics* 11, pp.105-112.

Muellbauer, J. [1974], "The Political Economy of Price Indices", Birkbeck Discussion Paper 22, March.

Pigou, A.C. [1920], *The Economics of Welfare*, London: Macmillan.

Pollak, R.A. [1969], "Conditional Demand Functions and Consumption Theory", *Quarterly Journal of Economics*, 83, pp.60-78.

_____ [1983], "The Theory of the Cost-of-Living Index," *Price Level Measurement: Proceedings From a Conference Sponsored by Statistics Canada.*

_____ [1975a], "Subindexes of the Cost-of-Living Index", *International Economic Review* 16, pp.135-150.

_____ [1975b], "The Intertemporal Cost-of-Living Index", *Annals of Economic and Social Measurement* 4, pp.179-196.

_____ [1981], "The Social Cost-of-Living Index", *Journal of Public Economics* 15, pp.311-336.

Prais, S. [1959], "Whose Cost-of-Living?", *Review of Economic Studies* 26, pp.126-134.

Riddell, W.C. [1983], "The Consumer Price Index and the Value of Leisure Time" *Price Level Measurement: Proceedings From a Conference Sponsored by Statistics Canada.*

Ruggles, R. [1967], "Price Indexes and International Price Comparisons", pp.171-205 in *Ten Economic Studies in The Tradition of Irving Fisher*, W. Fellner et al (eds.), New York: John Wiley.

Rymes, T.K. [1979], "The Treatment of Home Ownership in the CPI", *Review of Income and Wealth* 25, pp.393-412.

Samuelson, P.A. [1947], *Foundations of Economic Analysis*, Cambridge, Mass.: Harvard University Press.

_____ [1950], "Evaluation of Real National Income", *Oxford Economic Papers* 2, pp.1-29.

_____ [1974], "Complementarity - An Essay on the 40th Anniversary of the Hicks-Allen Revolution in Demand Theory", *The Journal of Economic Literature* 12, pp.1255-1289.

_____ and S. Swamy [1974], "Invariant Economic Index Numbers and Canonical Duality: Survey and Synthesis", *American Economic Review*, 64, pp.566-593.

Szulc, B.J. [1982], "Linking Index Number Series", *Price Level Measurement: Proceedings From a Conference Sponsored by Statistics Canada.*

Shephard, R.W. [1953], *Cost and Production Functions*, Princeton: Princeton University Press.

Statistics Canada [1982], ''The Consumer Price Index Reference Paper: Concepts and Procedures'', Catalogue No. 62-553, Occasional, Ottawa, Canada.

Triplett, J.E. [1982], ''Concepts of Quality in Input and Output Price Measures: A Resolution of the User Value-Resource Cost Debate'', in *The U.S. National Income and Product Accounts: Selected Topics*, M. Fuss (ed.), Chicago, The University of Chicago Press.

Usher, D. [1980], *The Measurement of Economic Growth*, New York; Columbia University Press.

Varian, H. [1982], ''The Nonparametric Approach to Demand Analysis'', *Econometrica*, 50, pp.945-973.

Vartia, Y.O. [1976], ''Ideal Log-Change Index Numbers'', *Scandinavian Journal of Statistics* 3, pp.121-126.

_____ [1978], ''Fisher's Five Tined Fork and other Quantum Theories of Index Numbers'', pp.271-295 in *Theory and Economic Applications of Economic Indices*, W. Eichhorn, R. Henn, O. Opitz and R.W. Shephard, (eds.), Wurzburg: Physica-Verlag.

PRICE LEVEL MEASUREMENT – W.E. Diewert (Editor)
Canadian Government Publishing Centre /
Elsevier Science Publishers B.V. (North-Holland)
© Minister of Supply and Services Canada, 1990

COMMENTS

R. Robert Russell
Department of Economics
New York University

Selling an economics paper is much like selling a consumer product; its sales prospects depend on packaging and promotion as well as content. My own casual empirical investigation suggests that there might even be an inverse correlation between content and promotion, much as inferior chiantis come in fancy straw baskets while the finer chiantis come in plain Bordeaux-type bottles.

When it comes to the ratio of packaging to content, Erwin Diewert's papers are at the extreme left-hand side of the spectrum – not only because the content is always so substantial, but also because he so eschews promotion.

Take, for example, Diewert's Theorem 1. To characterize this theorem, I draw your attention to the rather bifurcated research on index number theory in recent years. In the so-called mechanistic approach, represented at this conference in the paper by Wolfgang Eichhorn [1983], index numbers are not based on utility maximization; indeed, no assumptions are made about relationships between prices and quantities (i.e., about demand curves). The alternative approach, based on the theory of utility maximization, has a long history, outlined in many of the papers presented at this conference (especially Pollak [1983]). By providing necessary and sufficient conditions for a mechanistic price index to be rationalized by the theory of constrained cost minimization, Diewert's theorem synthesizes the two seemingly unrelated approaches to cost-of-living indexes. So how does Diewert characterize his results? As follows: "The above theorem is very closely related to some results of Pollak [1983] ...". Diewert's result is closely related to that of Pollak, but it is also much more.

All of Diewert's papers are densely packed with valuable results, and this one is no exception. This paper also evinces another important aspect of his papers: their high scholarly quality – not just, or even primarily, mathematical rigor, but also the careful attention that is paid to the history of economic thought. To a certain extent, this paper is a rigorous exercise in the history of economic thought, since many of the results are traced to the early work of, among others, Konüs and Malmquist. Especially important is the Konüs result on the existence of a reference utility level that bounds the true cost-of-living index between the Paasche and Laspeyres indexes. This result, which is used extensively by Diewert throughout this paper, is especially important when the Paasche and Laspeyres indexes are "close" to one another, in which case the true cost-of-living index for the intermediate reference utility level is fairly narrowly delineated. As Diewert puts it, "We will have the consumer's true cost-of-living between periods 0 and 1 for all practical purposes".

But enough accolades. In the remainder of this comment, I discuss one aspect of the paper that I believe is expositionally deficient. In particular, I attempt to shed some light on the relationship between two classes of indexes examined in Diewert's paper: The Konüs price index (the usual utility-based formula) and the Malmquist quantity index (which Diewert, with characteristically good taste, seems to favour among the several quantity indexes that he discusses).

In Diewert's presentation, these two indexes seem to be quite different types of animals – one based on constrained cost minimization and the other on distance functions. In fact, because of some important duality results that can be found in Shephard [1970] and Blackorby, Primont, and Russell [1978, Ch. 2], the Konüs and Malmquist indexes are completely symmetric and constitute a natural pair of quantity and price indexes. (This symmetry was first pointed out in Blackorby and Russell [1977].)

Diewert's (standard) derivation of the Konüs cost-of-living index is summarized in panel A of Table 1. The cost function, C, is derived by choosing a consumption vector, x, to minimize total expenditure subject to a utility constraint. The value, $C(u,p)$, can be written as the inner product of the price vector, p, and the vector-valued, constant-utility (Hicksian) demand function, $\zeta(u,p)$. This minimization problem is depicted in the diagram by the tangency of the equal-cost line (with normal $|\bar{p}|$) and the u indifference curve. The value of the cost function at $(\bar{u},\bar{p})$ is given by the horizontal intercept of the equal-cost line multiplied by $\bar{p}_1$. The Konüs cost-of-living index is then defined as the ratio of the values of the cost functions for the two different price vectors, p^0 and p^1.

TABLE 1

THE KONUS AND MALMQUIST INDICES CONSTRUCTED WITH COST FUNCTIONS

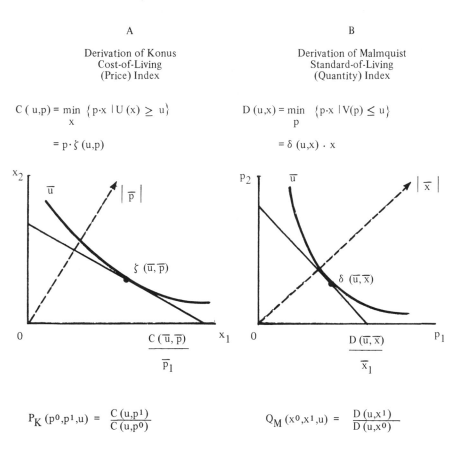

A	B
Derivation of Konus Cost-of-Living (Price) Index	Derivation of Malmquist Standard-of-Living (Quantity) Index

$$C(u,p) = \min_{x} \{p \cdot x \mid U(x) \ge u\}$$

$$= p \cdot \zeta(u,p)$$

$$D(u,x) = \min_{p} \{p \cdot x \mid V(p) \le u\}$$

$$= \delta(u,x) \cdot x$$

$$P_K(p^0, p^1, u) = \frac{C(u, p^1)}{C(u, p^0)}$$

$$Q_M(x^0, x^1, u) = \frac{D(u, x^1)}{D(u, x^0)}$$

Diewert's derivation of the Malmquist quantity index (called the standard-of-living index by Blackorby and Russell [1977]), is shown in panel B of Table 2. The distance function, D, with image $D(u,x)$, is defined as the maximal (proportional) amount by which x can be reduced and still be contained in the u upper level set. Thus, in the diagram, the distance function for utility level $\bar{u}$ and consumption vector $\bar{x}$ is given by the ratio α/β. The Malmquist quantity index is defined as the ratio of these distance functions for two different quantity vectors, x^0 and x^1.

These two concepts and their derivations seem quite disparate. Because of the duality between distance functions and cost functions, however, they are perfectly symmetric. The Malmquist quantity index can alternatively be derived by the cost-minimization procedure shown in panel B of Table 1. Here, the function D is defined as the minimum imputed value of a given quantity vector, subject to the constraint that the indirect utility function evaluated at the chosen (shadow) price vector be no greater than u. This (imputed) cost function is identical to the distance function in panel B of Table 2. It can be written as the inner product of the vector-valued, constant-utility imputed-price function, $\delta(u,x)$, and the quantity vector, x.

The cost-minimization problem is illustrated in panel B by the tangency between the plane with normal $|\bar{x}|$ and the indirect indifference curve labelled $\bar{u}$ at the price vector $\delta(\bar{u},\bar{x})$. The function value, $D(\bar{u},\bar{x})$, can then be obtained by multiplying the horizontal intercept of this plane by $\bar{x}_1$. The Malmquist quantity index is then defined as the ratio of values of the cost-imputation function at two different quantity vectors, x^0 and x^1. The juxtaposition of panels A and B of Table 1 make the complete symmetry between this derivation of the Malmquist quantity index and the standard derivation of the Konüs cost-of-living index apparent. The interchange of $(x,U,C,\geq)$ and $(p,V,D,\leq)$ yields equivalent derivations.

Just as the Malmquist index, defined as the ratio of distance functions by Diewert, can also be written as a ratio of cost-imputation functions, so can the Konüs cost-of-living index, traditionally defined as the ratio of cost functions, be defined as a ratio of distance functions. This is illustrated in panel A of Table 2. $C(u,p)$ is the maximum proportional amount by which the price vector, p, can be reduced while placing it in the u lower-level set in price space (recall that V is non-increasing in p). In the diagram, $C(\bar{u},\bar{p})$ is given by the ratio of the two line segments in the diagram, α/β. The Konüs cost-of-living index can be written as the ratio of values of the distance functions for two different price vectors, p^0 and p^1. The complete symmetry between these two indexes is revealed by comparing panels A and B of Table 2, where each is depicted as a ratio of values of a distance function. The interchange of $(x,U,D,\geq)$ and $(p,V,C,\leq)$ yields equivalent derivations.

To conclude, the Malmquist quantity index, which Diewert favours because of its intrinsic properties, is in fact the *only* natural counterpart to the widely accepted Konüs cost-of-living index.

TABLE 2

THE KONUS AND MALMQUIST INDICES CONSTRUCTED WITH DISTANCE FUNCTIONS

A	B
Derivation of Konus Cost-of-Living (Price) Index	Derivation of Malmquist Standard-of-Living (Quantity) Index

$$C(u,p) = \max_{\lambda} \ \{\lambda \mid V(p/\lambda) \le u\} \qquad\qquad D(u,x) = \max_{\lambda} \ \{\lambda \mid U(x/\lambda) \ge u\}$$

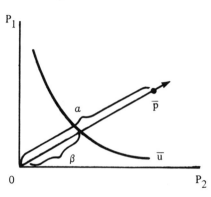 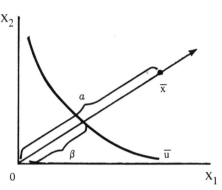

$$C(\bar{u},\bar{p}) = a/\beta \qquad\qquad\qquad D(\bar{u},\bar{x}) = a/\beta$$

$$P_K(p^0,p^1,u) = \frac{C(u,p^1)}{C(u,p^0)} \qquad\qquad Q_M(x^0,x^1,u) = \frac{D(u,x^1)}{D(u,x^0)}$$

References

Blackorby, C., D. Primont and R.R. Russell [1978], *Duality Separability, and Functional Structure; Theory and Economic Applications*, New York: North-Holland.

_____ and R.R. Russell [1977], "Indices and Subindices of the Cost of Living and the Standard of Living", *International Economic Review* 18, pp.229-240.

Eichhorn, W. [1983], "The Axiomatic Foundations of Price Indices and Purchasing Power Parities", *Price Level Measurement: Proceedings From a Conference Sponsored by Statistics Canada.*

Pollak, R.A. [1983], "The Theory of the Cost-of-Living Index", *Price Level Measurement: Proceedings From a Conference Sponsored by Statistics Canada.*

Shephard, R. [1970], *The Theory of Cost and Production Functions*, Princeton: Princeton University Press.

PRICE LEVEL MEASUREMENT – W.E. Diewert (Editor)
Canadian Government Publishing Centre /
Elsevier Science Publishers B.V. (North-Holland)
© Minister of Supply and Services Canada, 1990

INDIVIDUAL AND SOCIAL COST-OF-LIVING INDEXES

Dale W. Jorgenson
Department of Economics
Harvard University
Daniel T. Slesnick
Department of Economics
University of Texas

SUMMARY

The purpose of this paper is to present an econometric approach to cost-of-living measurement. This approach implements the economic theory of individual cost-of-living measurement pioneered by Konüs [1939] almost six decades ago. In this paper we develop and implement a completely parallel theory of social cost-of-living measurement.

Our approach to cost-of-living measurement is based on an econometric model of aggregate consumer behavior. The novel feature of this model is that systems of individual demand functions can be recovered uniquely from the system of aggregate demand functions. We derive cost-of-living indexes for individual households from systems of individual demand functions.

Our key innovation in the economic theory of cost-of-living measurement is the introduction of an explicit social welfare function. Our social welfare function incorporates measures of individual welfare from our econometric model. In addition, this social welfare function employs normative criteria for evaluating transfers among individuals.

Given measures of individual welfare from our econometric model, we can express the level of social welfare as a function of prices and of total expenditures of all consuming units. We present methods for translating changes in prices into measures of change in the social cost-of-living. Our definitions of social and individual cost-of-living indexes are perfectly analogous.

Finally, we extend the concept of a social cost-of-living index to groups of consuming units with common demographic characteristics. Our definition of the group cost-of-living index is analogous to the definition of a social cost-of-living index. To implement a group cost-of-living index we require a group welfare function that is analogous to a social welfare function.

1. Introduction

The purpose of this paper is to present an econometric approach to cost-of-living measurement. This approach implements the economic theory of individual cost-of-living measurement pioneered by Konüs [1939] almost six decades ago.[1] In this paper we develop and implement a completely parallel theory of social cost-of-living measurement.

Our approach to cost-of-living measurement is based on an econometric model of aggregate consumer behavior. The novel feature of this model is that systems of individual demand functions can be recovered uniquely from the system of aggregate demand functions. We derive cost-of-living indexes for individual households from systems of individual demand functions.

Our key innovation in the economic theory of cost-of-living measurement is the introduction of an explicit social welfare function. Our social welfare function incorporates measures of individual welfare from our econometric model. In addition, this social welfare function employs normative criteria for evaluating transfers among individuals.

In Section 2 we outline econometric methodology for developing a model of aggregate consumer behavior. In this model the system of aggregate demand functions depends on summary statistics of the joint distribution of attributes and total expenditures of individual households. Attributes of households such as demographic characteristics enable us to account for differences in preferences.

In Section 3 we implement our econometric model of aggregate consumer behavior for the United States. For this purpose we employ cross-section data on individual expenditure patterns. We combine these data with time series information on aggregate expenditure patterns. We also employ time series data on the distribution of total expenditures among consuming units.

In Section 4 we present methods for translating changes in prices into measures of change in the individual cost of living. For this purpose we employ the individual expenditure function. The expenditure function gives the minimum total expenditure required to attain a base level of individual welfare. This minimum expenditure depends on prices and on the attributes of the consuming unit.

Following Konüs [1939], we define the individual cost-of-living index as the ratio between the total expenditure required to attain a base level of individual welfare at current prices and the corresponding base level of expenditure. If the individual cost-of-living index exceeds unity and total expenditure is constant, then the welfare of the consuming unit has decreased relative to the base level.

To implement the individual cost-of-living index the remaining problem is to determine the base level of individual welfare. For this purpose we define individual welfare in terms of the indirect utility function. This function gives utility as a function of prices and of the total expenditure and attributes of the consuming unit.

We calculate individual cost-of-living indexes and rates of inflation for households with different levels of total expenditure and different demographic characteristics for the period 1958-1978. For this purpose we derive indirect utility functions and expenditure functions for all consuming units from our econometric model of aggregate consumer behavior.

In Section 5 we present methods for evaluating the level of social welfare. For this purpose we construct an explicit social welfare function. Our social welfare function incorporates measures of individual welfare based on indirect utility functions for all consuming units. In addition, this social welfare function employs normative criteria based on horizontal and vertical equity for evaluating transfers among units.

Given indirect utility functions from our econometric model, we can express the level of social welfare as a function of prices and of total expenditures and attributes for all consuming units. We define the social expenditure function as the minimum aggregate expenditure required to attain a base level of social welfare. This minimum level of expenditure depends on prices and on the attributes of all consuming units.

In Section 6 we present methods for translating changes in prices into measures of change in the social cost of living. Following Pollak [1981], we define the social cost-of-living index as the ratio between the aggregate expenditure required to attain a base level of social welfare at current prices and the corresponding base level of expenditure. If the social cost-of-living index exceeds unity and aggregate expenditure is constant, then social welfare has decreased relative to the base level.

Our definitions of social and individual cost-of-living indexes are perfectly analogous. In these definitions the roles of the social and individual expenditure functions and the roles of the social and individual welfare functions are completely parallel. We calculate the social cost-of-living index and rates of inflation for the period 1958-1978.

In Section 7 we extend the concept of a social cost-of-living index to groups of consuming units with common demographic characteristics. Our definition of the group cost-of-living index is analogous to the definition of a social cost-of-living index. To implement a group cost-of-living index we require a group welfare function that is analogous to a social welfare function. We also require a group expenditure function.

Given group welfare and expenditure functions, we calculate cost-of- living indexes and rates of inflation for groups of households with common demographic characteristics for the period 1958-1978. If a group cost-of- living index exceeds unity and group expenditure is constant, then the welfare of the group has decreased relative to the base level.

In Section 8 we compare the econometric and index number approaches to cost-of-living measurement. The econometric approach incorporates all the information employed in cost-of-living index numbers. An important advantage of the econometric approach is that it summarizes the available information in a concise and readily intelligible way.

In concluding we emphasize that the econometric and index number approaches share a number of significant limitations. These limitations arise from the practical problems of obtaining appropriate data on prices and expenditures. However, the econometric approach has greater flexibility and is easier to apply. These advantages are illustrated by our implementation of individual, social and group cost-of-living indexes for the United States for the period 1958-1978.

2. Aggregate Consumer Behavior

In this section we develop an econometric model of aggregate consumer behavior based on the theory of exact aggregation, following Jorgenson, Lau and Stoker [1980, 1981, 1982]. Our model incorporates time series data on prices and aggregate quantities consumed. We also include cross-section data on individual quantities consumed, individual total expenditure, and attributes of individual households such as demographic characteristics.

To construct an econometric model based on exact aggregation we first represent individual preferences by means of an indirect utility function for each consuming unit, using the following notation:

p_n – price of the nth commodity, assumed to be the same for all consuming units.

$p = (p_1, p_2 \dots p_N)$ – the vector of prices of all commodities.

x_{nk} – the quantity of the nth commodity group consumed by the kth consuming unit $(n = 1, 2 \ldots N; k = 1, 2 \ldots K)$.

$$M_k = \sum_{n=1}^{N} p_n x_{nk} \text{ – total expenditure of the kth consuming unit } (k = 1, 2 \ldots K).$$

$w_k = p_n x_{nk} / M_k$ – expenditure share of the nth commodity group in the budget of the kth consuming unit $(n = 1, 2 \ldots N; k = 1, 2 \ldots K)$.

$w_k = (w_{1k}, w_{2k} \ldots w_{Nk})$ – vector of expenditure shares for the kth consuming unit $(k = 1, 2 \ldots K)$.

$$\ln \frac{p}{M_k} = (\ln \frac{p_1}{M_k}, \ln \frac{p_2}{M_k} \ldots \ln \frac{p_N}{M_k}) \text{ – vector of logarithms of ratios of prices to expenditure}$$

by the kth consuming unit $(k = 1, 2 \ldots K)$.

$\ln p = (\ln p_1, \ln p_2 \ldots \ln p_N)$ – vector of logarithms of prices.

A_k – vector of attributes of the kth consuming unit $(k = 1, 2 \ldots K)$.

We assume that the kth consuming unit allocates expenditures in accord with the transcendental logarithmic or translog indirect utility function,[2] say V_k, where:

$$\ln V_k = G(\ln \frac{p}{M_k}'\alpha_p + \frac{1}{2} \ln \frac{p}{M_k}'B_{pp}\ln \frac{p}{M_k} + \ln \frac{p}{M_k}'B_{pA}A_k, A_k), \quad (k = 1, 2 \ldots K). \tag{2.1}$$

In this representation the function G is a monotone increasing function of the variable

$\ln \frac{p}{M_k}'\alpha_p + \frac{1}{2}\ln \frac{p}{M_k}'B_{pp}\ln \frac{p}{M_k} + \ln \frac{p}{M_k}'B_{pA}A_k$. In addition, the function G depends directly on the attribute vector A_k.[3] The vector α_p and the matrices B_{pp} and B_{pA} are constant parameters that are the same for all consuming units.

The expenditure shares of the kth consuming unit can be derived by the logarithmic form of Roy's [1943] Identity:[4]

$$w_{nk} = \frac{\partial \ln V_k}{\partial \ln (p_n/M_k)} \Big/ \sum_{n=1}^{N} \frac{\partial \ln V_k}{\partial \ln (p_n/M_k)}, \quad (n = 1, 2...N; \ k = 1, 2 \ldots K). \qquad (2.2)$$

Applying this Identity to the translog indirect utility function (2.1), we obtain the system of individual expenditure shares:

$$w_k = \frac{1}{D_k(p)} \left(\alpha_p + B_{pp} \ln \frac{p}{M_k} + B_{pA} A_k \right), \quad (k = 1, 2 \ldots K), \qquad (2.3)$$

where the denominators $\{D_k\}$ take the form:

$$D_k = i'\alpha_p + i'B_{pp} \ln \frac{p}{M_k} + i'B_{pA} A_k, \quad (k = 1, 2 \ldots K). \qquad (2.4)$$

The individual expenditure shares are homogeneous of degree zero in the unknown parameters -- α_p, B_{pp}, B_{pA}. By multiplying a given set of these parameters by a constant we obtain another set of parameters that generates the same system of individual budget shares. Accordingly, we can choose a normalization for the parameters without affecting observed patterns of individual expenditure allocation. We find it convenient to employ the normalization:

$$i'\alpha_p = -1.$$

Under this restriction any change in the set of unknown parameters will be reflected in changes in individual expenditure patterns.

The conditions for exact aggregation are that the individual expenditure shares are linear in functions of the attributes $\{A_k\}$ and total expenditures $\{M_k\}$ for all consuming units[5]. These conditions will be satisfied if and only if the terms involving the attributes and expenditures do not appear in the denominators of the expressions given above for the individual expenditure shares, so that:

$$i' B_{pp} i = 0,$$

$$i' B_{pA} = 0.$$

The exact aggregation restrictions imply that the denominators $\{D_k\}$ reduce to:

$$D = -1 + i' B_{pp} \ln p,$$

where the subscript k is no longer required, since the denominator is the same for all consuming units. Under these restrictions the individual expenditure shares can be written:

$$w_k = \frac{1}{D(p)} (\alpha_p + B_{pp} \ln p - B_{pp} i \cdot \ln M_k + B_{pA} A_k), \quad (k = 1, 2, \ldots K). \tag{2.5}$$

The individual expenditure shares are linear in the logarithms of expenditures $\{\ln M_k\}$ and in the attributes $\{A_k\}$, as required by exact aggregation.

Under exact aggregation the indirect utility function for each consuming unit can be represented in the form:

$$\ln V_k = F(A_k) + \ln p' (\alpha_p + \frac{1}{2} B_{pp} \ln p + B_{pA} A_k) - D(p) \ln M_k, \quad (k = 1, 2, \ldots K). \tag{2.6}$$

In this representation the indirect utility function is linear in the logarithm of total expenditure $\ln M_k$ with a coefficient that depends on the prices p (k = 1, 2 ... K). This property is invariant with respect to positive affine transformations, but is not preserved by arbitrary monotone increasing transformations. We conclude that the indirect utility function (2.6) provides a cardinal measure of utility for each consuming unit.

If a system of individual expenditure shares (2.3) can be generated from an indirect utility function of the form (2.1) we say that the system is **integrable**. A complete set of conditions for integrability[6] is the following:

1. Homogeneity. The individual expenditure shares are homogeneous of degree zero in prices and total expenditure.

We can also write the individual expenditure shares in the form:

$$\beta_{pM} = B_{pp}i.$$ (2.7)

Given the exact aggregation restrictions, there are N-1 restrictions implied by homogeneity.

2. Summability. The sum of the individual expenditure shares over all commodity groups is equal to unity:

$$i'w_k = 1, \qquad (k = 1, 2 \dots K).$$

We can write the denominator $D(p)$ in (2.4) in the form:

$$D = -1 + \beta_{Mp}\ln p,$$

where the vector of parameters β_{Mp} is constant and the same for all commodity groups and all consuming units. Summability implies that this vector must satisfy all restrictions:

$$\beta_{Mp} = i'B_{pp}.$$ (2.8)

Given the exact aggregation restrictions, there are N-1 restrictions implied by summability.

3. Symmetry. The matrix of compensated own- and cross-price substitution effects must be symmetric.

If the system of individual expenditure shares can be generated from an indirect utility function of the form (2.1), a necessary and sufficient condition for symmetry is that the matrix B_{pp} must be symmetric. Without imposing this condition, we can write the individual expenditure shares in the form:

$$w_k = \frac{1}{D(p)} (\alpha_p + B_{pp} \ln \frac{p}{M_k} + B_{pA} A_k), \quad (k = 1, 2 \ldots K).$$

Symmetry implies that the matrix of parameters B_{pp} must satisfy the restrictions:

$$B_{pp} = B'_{pp} \quad . \tag{2.9}$$

The total number of symmetry restrictions if $\frac{1}{2}N(N-1)$.

4. Nonnegativity. The individual expenditure shares must be nonnegative.

By summability the individual expenditure shares sum to unity, so that we can write:

$$w_k \geq 0, \qquad (k = 1, 2 \ldots K),$$

where $w_k \geq 0$ implies $w_{nk} \geq 0$, $(n = 1, 2 \ldots N)$, and $w_k \neq 0$.

Since the translog indirect utility function is quadratic in the logarithms of prices, we can always choose the prices so that the individual expenditure shares violate the nonnegativity conditions. Accordingly, we cannot impose restrictions on the parameters of the translog indirect utility functions that would imply nonnegativity of the individual expenditure shares. Instead we consider restrictions on the parameters that imply monotonicity of the system of individual demand functions for all nonnegative expenditure shares.

5. Monotonicity. The matrix of compensated own- and cross-price substitution effects must be nonpositive definite.

We introduce the definition due to Martos [1969] of a **strictly merely positive subdefinite**

matrix, namely, a real symmetric matrix S such that:

$$xSx < 0$$

implies $Sx > 0$ or $Sx < 0$. A necessary and sufficient condition for monotonicity is either that the translog indirect utility function is homothetic or that B_{pp}^{-1} exists and is strictly merely positive subdefinite.[7]

To provide a basis for evaluating the impact of transfers among households on social welfare, we find it useful to represent household preferences by means of a utility function that is the same for all consuming units. For this purpose, we assume that the kth consuming unit maximizes its utility, say U_k, where:

$$U_k = U \left[\frac{x_{1k}}{m_1 (A_k)}, \frac{x_{2k}}{m_2 (A_k)} \cdots \frac{x_{Nk}}{m_N (A_k)} \right], \quad (k = 1, 2 \ldots K), \tag{2.10}$$

subject to the budget constraint:

$$M_k = \sum_{n=1}^{N} p_n x_{nk}, \quad (k = 1, 2 \ldots K).$$

In this representation of consumer preferences the quantities $\{x_{nk}/m_n(A_k)\}$ can be regarded as **effective quantities consumed**, as proposed by Barten [1964]. The crucial assumption embodied in this representation is that differences in preferences among consumers enter the utility function U only through differences in the commodity specific household equivalence scales $\{m_n(A_k)\}$.[8]

Consumer equilibrium implies the existence of an indirect utility function, say V, that is the same for all consuming units. The level of utility for the kth consuming unit, say V_k, depends on the prices of individual commodities, the household equivalence scales, and the level of total expenditure:

$$
V_k = V \left[\frac{p_1 m_1 (A_k)}{M_k}, \frac{p_2 m_2 (A_k)}{M_k} \cdots \frac{p_N m_N (A_k)}{M_k} \right], \quad (k = 1, 2 \ldots K). \tag{2.11}
$$

In this representation the prices $\{p_n m_n(A_k)\}$ can be regarded as **effective prices**. Differences in preferences among consuming units enter this indirect utility function only through the household equivalence scales $\{m_n(A_k)\}$ $(k = 1, 2 \ldots K)$.

To represent the translog indirect utility function (2.1) in terms of household equivalence scales, we require some additional notation:

$$
\ln \frac{p \, m(A_k)}{M_k} \text{ – vector of logarithms of ratios of effective prices}
$$

$\{p_n \, m_n(A_k)\}$ to total expenditure M_k of the kth consuming unit $(k = 1, 2 \ldots K)$.

$\ln m(A_k) = (\ln m_1(A_k), \ln m_2(A_k) \ldots \ln m_N(A_k))$ – vector of logarithms of the household equivalence scales of the kth consuming unit $(k = 1, 2 \ldots K)$.

We assume, as before, that the kth consuming unit allocates its expenditures in accord with the translog indirect utility function (2.1). However, we also assume that this function, expressed in terms of the effective prices $\{p_n \, m_n(A_k)\}$ and total expenditure M_k, is the same for all consuming units. The indirect utility function takes the form:

$$
\ln V_k = \ln \frac{p \, m(A_k)}{M_k} {}'\alpha_p + \frac{1}{2} \ln \frac{p \, m(A_k)}{M_k} {}' B_{pp} \ln \frac{p \, m(A_k)}{M_k}, \quad (k = 1, 2 \ldots K). \tag{2.12}
$$

Taking logarithms of the effective prices $\{p_n \, m_n(A_k)\}$, we can rewrite the indirect utility function (2.12) in the form:

$$\ln V_k = \ln m(A_k)' \alpha_p + \frac{1}{2} \ln m(A_k)' B_{pp} \ln m(A_k) + \ln \frac{p}{M_k}' \alpha_p \qquad (2.13)$$

$$+ \frac{1}{2} \ln \frac{p}{M_k}' B_{pp} \ln \frac{p}{M_k} + \ln \frac{p}{M_k}' B_{pp} \ln m(A_k), \qquad (k = 1, 2 \dots K).$$

Comparing the representation (2.13) with the representation (2.6), we see that the term involving only the household equivalent scales must take the form:

$$F(A_k) = \ln m(A_k)' \, \alpha_p + \frac{1}{2} \ln m(A_k)' B_{pp} \ln m(A_k), \, (k = 1, 2 \dots K). \qquad (2.14)$$

Second, the term involving ratios of prices to total expenditure and the household equivalence scales must satisfy:

$$\ln \frac{p}{M_k}' B_{pA} A_k = \ln \frac{p}{M_k}' B_{pp} \ln m(A_k), \qquad (k = 1, 2 \dots K). \qquad (2.15)$$

for all prices and total expenditure.

The household equivalence scales $\{m_n(A_k)\}$ defined by (2.15) must satisfy the equation:

$$B_{pA} A_k = B_{pp} \ln m(A_k), \qquad (k = 1, 2 \dots K). \qquad (2.16)$$

Under monotonicity of the individual expenditure shares the matrix B_{pp} has an inverse, so that we can express the household equivalence scales in terms of the parameters of the translog indirect utility function – B_{pp} , B_{pA} – and the attributes $\{A_k\}$:

$$\ln m(A_k) = B_{pp}^{-1} B_{pA} A_k , \qquad (k = 1, 2 \dots K). \qquad (2.17)$$

We can refer to these scales as the **commodity specific translog household equivalence scales**.

Substituting the commodity specific equivalence scales (2.16) into the indirect utility function (2.13) we obtain a representation of the indirect utility function in terms of the attributes $\{A_k\}$:

$$\ln V_k = A'_k B'_{pA} B_{pp}^{-1} \alpha_p + \frac{1}{2} A'_k B'_{pA} B_{pp}^{-1} B_{pA} A_k \tag{2.18}$$

$$+ \ln p' (\alpha_p + \frac{1}{2} B_{pp} \ln p + B_{pA} A_k) - D(p)\ln M_k , \quad (k = 1, 2 \dots K).$$

This form of the translog indirect utility function is equivalent to the form (2.1) in that both generate the same system of individual demand functions. By requiring that the attributes A_k enter only through the commodity specific household equivalence scales, we have provided a specific form for the function $F(A_k)$ in (2.6).

Given the indirect utility function (2.18) for each consuming unit, we can express total expenditure as a function of prices, consumer attributes, and the level of utility:

$$\ln M_k = \frac{1}{D(p)} [A'_k B'_{pA} B_{pp}^{-1} \alpha_p + \frac{1}{2} A'_k B'_{pA} B_{pp}^{-1} B_{pA} A_k \tag{2.19}$$

$$+ \ln p'(\alpha_p + \frac{1}{2} B_{pp} \ln p + B_{pA} A_k) - \ln V_k], \quad (k = 1, 2 \dots K).$$

We can refer to this function as the **translog expenditure function**. The translog expenditure function gives the minimum expenditure required for the kth consuming unit to achieve the utility level V_k, given prices p (k = 1, 2 ... K).

We find it useful to introduce household equivalence scales that are not specific to a given commodity.[9] Following Deaton and Muellbauer [1980], we define a general household equivalence scale, say m_0, as follows:

$$m_0 = \frac{M_k [p \, m \, (A_k), V_k^0]}{M_0(p, V_k^0)}, \quad (k = 1, 2 \dots K). \tag{2.20}$$

where M_k is the expenditure function for the kth household, M_0 is the expenditure function for a reference household with commodity specific equivalence scales equal to unity for all commodities, and $p\, m(A_k)$ is a vector of effective prices $\{p_n\, m_n(A_k)\}$.

The general household equivalence scale m_0 is the ratio between total expenditure required by the kth household and by the reference household for the same level of utility V_k^0 (k = 1, 2 ... K). This scale can be interpreted as the number of household equivalent members. The number of members depends on the attributes A_k of the consuming unit and on the prices p.

If each household has a translog indirect utility function, then the general household equivalence scale for the kth household takes the form:

$$\ln m_0 = \ln M_k - \ln M_0 , \qquad (2.21)$$

$$= \frac{1}{D(p)} [\ln m\, (A_k)'\alpha_p + \frac{1}{2}\ln m\, (A_k)'B_{pp}\ln m\, (A_k) + \ln m\, (A_k)'B_{pp}\ln p],$$

$$(k = 1, 2 ... K).$$

We can refer to this scale as the **general translog household equivalence scale**. The translog equivalence scale depends on the attributes A_k of the kth household and the prices p of all commodities, but is independent of the level of utility V_k^0.

Given the general translog equivalence scale, we can rewrite the indirect utility function (2.18) in the form:

$$\ln V_k = \ln p'\alpha_p + \frac{1}{2}\ln p'B_{pp}\ln p - D(p) \ln [M_k/m_0(p, A_k)], (k = 1, 2 ... K). \qquad (2.22)$$

The level of utility for the kth consuming unit depends on prices p and total expenditure per household equivalent member $M_k/m_0(p, A_k)$ (k = 1, 2 ... K). Similarly, we can rewrite the expenditure function (2.19) in the form:

$$\ln M_k = \frac{1}{D(p)} [\ln p'(\alpha_p + \frac{1}{2}B_{pp}\ln p) - \ln V_k] + \ln m_0(p, A_k),$$

$$(k = 1, 2 \ldots K). \tag{2.23}$$

Total expenditure required by the kth consuming unit to attain the level of utility V_k depends on prices p and the number of household equivalent members $m_0(p, A_k)$ (k = 1, 2 ... K).

To construct an econometric model of aggregate consumer behavior based on exact aggregation we obtain aggregate expenditure shares, say w, by multiplying individual expenditure shares (2.5) by expenditure for each consuming unit, adding over all consuming units, and dividing by aggregate expenditure, $M = \sum_{k=1}^{K} M_k$:

$$w = \frac{\sum M_k w_k}{M}. \tag{2.24}$$

The aggregate expenditure shares can be written:

$$w = \frac{1}{D(p)} (\alpha_p + B_{pp} \ln p - B_{pp}i \frac{\sum M_k \ln M_k}{M} + B_{pA} \frac{\sum M_k A_k}{M}). \tag{2.25}$$

The aggregate expenditure patterns depend on the distribution of expenditure over all consuming units through summary statistics of the joint distribution of expenditures and attributes – $\sum M_k \ln M_k / M$ and $\{\sum M_k A_k / M\}$. Systems of individual expenditure shares (2.5) for consuming units with identical demographic characteristics can be recovered in one and only one way from the system of aggregate expenditure shares (2.25).

To summarize: Systems of individual expenditure shares (2.5) can be recovered in one and only one way from the system of aggregate expenditure shares (2.25). Given a system of individual exenditure shares (2.5) that is integrable, we can recover the indirect utility function (2.22). This indirect utility function provides a cardinal measure of utility. We obtain measures of utility for all consuming units by deriving indirect utility functions from the fitted systems of individual expenditure shares.

3. Econometric Model

In this section we present the empirical results of implementing the econometric model of consumer behavior described in Section 2. We divide consumer expenditures among five commodity groups:

1. Energy: Expenditures on electricity, natural gas, heating oil and gasoline.

2. Food: Expenditures on all food products, including tobacco and alcohol.

3. Consumer Goods: Expenditures on all other nondurable goods included in consumer expenditures.

4. Capital Services: The service flow from consumer durables and the service flow from housing.

5. Consumer Services: Expenditures on consumer services, such as car repairs, medical services, entertainment, and so on.

We employ the following demographic characteristics as attributes of individual households:

1. Family size: 1, 2, 3, 4, 5, 6 and 7 or more persons.

2. Age of head: 16-24, 25-34, 35-44, 45-54, 55-64, 65 and over.

3. Region of residence: Northeast, North Central, South and West.

4. Race: White, nonwhite.

5. Type of residence: Urban, rural.

Our cross-section observations on individual expenditures for each commodity group and on demographic characteristics of individual households are for the year 1972 from the 1972-1973 Survey of Consumer Expenditures (CES).[10] Our time series observations are based on data on personal consumption expenditures from the United States National Income and Product Accounts (NIPA) for the years 1958 to 1974.[11] Prices for each commodity group are defined in terms of translog price indexes computed from detailed prices included in NIPA for each year. We employ time series data on the distribution of expenditures over all households and among demographic groups based on **Current Population Reports**.[12]

In our application we treat the expenditure shares for five commodity groups as endogenous variables, so that we estimate four equations. As unknown parameters we have four elements of the vector α_p, four expenditure coefficients of the vector $B_{pp}i$, 16 attribute coefficients for each of the four equations in the matrix B_{pA}, and 10 price coefficients in the matrix B_{pp}, which is constrained to be symmetric. The expenditure coefficients are sums of price coefficients in the corresponding equation, so that we have a total of 82 unknown parameters. We estimate the complete model, subject to inequality restrictions implied by monotonicity of the individual expenditure shares, by pooling time series and cross-section data.[13] The results are given in Table 1.

The impacts of changes in total expenditures and in demographic characteristics of the individual household are estimated very precisely. This reflects the fact that estimates of the expenditure and demographic effects incorporate a relatively large number of cross-section observations. The impacts of prices enter through the denominator of the equations for expenditure shares; these price coefficients are estimated very precisely since they also incorporate cross-section data. Finally, the price impacts also enter through the numerators of equations for the expenditure shares. These parameters are estimated somewhat less precisely, since they are based on a much smaller number of time series observations on prices.

To summarize: We have implemented an econometric model of aggregate consumer behavior by combining time series and cross-section data for the United States. This model allocates personal consumption expenditures among five commodity groups -- energy, food, other consumer goods, capital services, and other consumer services. Households are classified by five sets of demographic characteristics -- family size, age of head, region of residence, race, and urban versus rural residence.

TABLE 1. Pooled Estimation Results

Notation:

CONST   =   constant term.

ln PEN   =   coefficient of log of price of energy.

ln PF   =   coefficient of log of price of food.

ln PCG   =   coefficient of log of price of consumer goods.

ln PK   =   coefficient of log of price of capital services.

ln PCS   =   coefficient of log of price of consumer services.

ln M   =   coefficient of log of total expenditure.

S2   =   coefficient of dummy for family of size 2.

S3   =   coefficient of dummy for family of size 3.

S4   =   coefficient of dummy for family of size 4.

S5   =   coefficient of dummy for family of size 5.

S6   =   coefficient of dummy for family of size 6.

S7+   =   coefficient of dummy for family of size 7 or more.

A25-34   =   coefficient of dummy for age between 25 and 34.

A35-44   =   coefficient of dummy for age between 35 and 44.

A45-54   =   coefficient of dummy for age between 45 and 54.

A55-64   =   coefficient of dummy for age between 55 and 64.

A65+   =   coefficient of dummy for age 65 and over.

RNC   =   coefficient of dummy for family living in North Central.

RS   =   coefficient of dummy for family living in South.

RW   =   coefficient of dummy for family living in West.

NW   =   coefficient of dummy for nonwhite family.

RUR   =   coefficient of dummy for family living in rural area.

$$D(p) = {}^= 1 - .03491 \ln PEN - .08171 \ln PF + .06189 \ln PCG$$
$$\quad\quad (.000997) \quad\quad (.00238) \quad\quad (.00214)$$
$$\quad - .002060 \ln PK + .05679 \ln PCS$$
$$\quad\quad (.00300) \quad\quad (.00233)$$

TABLE 1. Energy – Continued

Parameter	Estimate	Standard Error
CONST	– .3754	.00923
ln PEN	.09151	.0134
ln PF	– .1441	.0214
ln PCG	– .06455	.0127
ln PK	.07922	.0171
ln PCS	.003061	.0138
ln M	.03491	.000997
S2	– .02402	.00139
S3	– .02971	.00163
S4	– .03144	.00178
S5	– .03255	.00206
S6	– .03606	.00249
S7 +	– .02977	.00266
A25-34	.0002010	.00197
A35-44	– .006703	.00210
A45-54	– .01155	.00199
A55-64	– .01372	.00199
A65 +	– .005487	.00196
RNC	– .003277	.00131
RS	.0001280	.00131
RW	.01281	.00140
NW	.01300	.00170
RUR	– .03057	.00134

TABLE 1. Food – Continued

Parameter	Estimate	Standard Error
CONST	− .8917	.0215
ln PEN	− .1441	.0214
ln PF	.3118	.0428
ln PCG	.05547	.0215
ln PK	− .1982	.0334
ln PCS	− .1066	.0259
ln M	.08171	.00238
S2	− .04859	.00333
S3	− .06730	.00390
S4	− .08881	.00428
S5	− .1108	.00496
S6	− .1185	.00598
S7 +	− .1471	.00639
A25-34	− .04393	.00474
A35-44	− .08221	.00504
A45-54	− .09604	.00478
A55-64	− .1034	.00477
A65 +	− .08833	.00470
RNC	.08173	.00315
RS	.01213	.00314
RW	.01856	.00337
NW	.006274	.00409
RUR	− .001793	.00323

TABLE 1. Consumer Goods – Continued

Parameter	Estimate	Standard Error
CONST	.4053	.0194
ln PEN	− .06455	.0127
ln PF	.05547	.0215
ln PCG	.2301	.0269
ln PK	− .1056	.0195
ln PCS	− .05354	.0271
ln M	− .06189	.00214
S2	− .005594	.00300
S3	− .006290	.00351
S4	− .001941	.00385
S5	.004522	.00446
S6	.01059	.00539
S7 +	.01495	.00575
A25-34	− .02311	.00426
A35-44	− .01916	.00454
A45-54	− .005279	.00431
A55-64	− .009068	.00429
A65 +	− .01722	.00423
RNC	− .02098	.00283
RS	− .03553	.00283
RW	− .009928	.00304
NW	− .02648	.00368
RUR	− .01122	.00290

TABLE 1. Capital Services – Continued

Parameter	Estimate	Standard Error
CONST	− .4658	.0270
ln PEN	.07922	.0171
ln PF	− .1982	.0334
ln PCG	− .1056	.0195
ln PK	.2038	.0368
ln PCS	.01869	.0165
ln M	.002060	.00300
S2	.07355	.00421
S3	.09982	.00493
S4	.1148	.00541
S5	.1253	.00626
S6	.1284	.00756
S7 +	.1369	.00807
A25-34	.04362	.00599
A35-44	.08503	.00637
A45-54	.1166	.00605
A55-64	.1395	.00603
A65 +	.1296	.00595
RNC	.02767	.00398
RS	.05528	.00397
RW	− .004132	.00427
NW	− .003539	.00517
RUR	.05588	.00408

TABLE 1. Consumer Services – Concluded

Parameter	Estimate	Standard Error
CONST	.3277	.0211
ln PEN	.003061	.0138
ln PF	− .1066	.0259
ln PCG	− .05354	.0271
ln PK	.01869	.0165
ln PCS	.1952	.0375
ln M	− .05679	.00233
S2	.004666	.00327
S3	.003483	.00383
S4	.007338	.00420
S5	.01357	.00486
S6	.01561	.00587
S7 +	.02508	.00627
A25-34	.02321	.00465
A35-44	.02304	.00495
A45-54	− .003805	.00470
A55-64	− .01332	.00468
A65 +	− .01863	.00462
RNC	− .02214	.00309
RS	− .03200	.00308
RW	− .01731	.00331
NW	.01074	.00401
RUR	− .01229	.00317

Convergence after 3 iterations.

SSR = 37387.12.

Convergence criterion = .00001.

4. Individual Cost-of-Living Indexes

In this section we outline a methodology for translating changes in prices into measures of change in the individual cost of living. The first step in measuring the cost of living for an individual consuming unit is to select a representation for the individual welfare function. We assume that individual welfare for the kth consuming unit, say W_k ($k = 1, 2 \ldots K$), is equal to the logarithm of the translog indirect utility function (2.22):

$$W_k = \ln V_k , \tag{4.1}$$

$$= \ln p' \alpha_p + \frac{1}{2} \ln p' B_{pp} \ln p - D(p) \ln [M_k / m_0(p, A_k)], \quad (k = 1, 2 \ldots K).$$

Following Konüs [1939], we define the cost-of-living index for the kth consuming unit, say P_k ($k = 1, 2 \ldots K$), as the ratio of two levels of total expenditure:

$$P_k(p^1, p^0, W_k^0, A_k) = \frac{M_k(p^1, W_k^0, A_k)}{M_k(p^0, W_k^0, A_k)} , \quad (k = 1, 2 \ldots K). \tag{4.2}$$

In this ratio the numerator is the expenditure required to attain the base level of individual welfare W_k^0 at the current price system p^1; the denominator is the base level of expenditure M_k^0.

The translog indirect utility function (2.22) and the translog expenditure function (2.23) can be employed in implementing the individual cost-of-living index (4.2).[14] First, we can express the base level of individual welfare W_k^0 in terms of the translog indirect utility function:

$$W_k^0 = \ln p^{0'} \alpha_p + \frac{1}{2} \ln p^{0'} B_{pp} \ln p^0 - D(p^0) \ln [M_k^0 / m_0(p^0, A_k)], \quad (k = 1, 2 \ldots K). \tag{4.3}$$

In this expression $m_0(p^0, A_k)$ is the base level of the general translog household equivalence scale (2.21) and M_k^0 is the base level of total expenditure.

Using the translog expenditure function, we can express the individual cost-of-living index (4.2) in the form:

$$\ln P_k(p^1, p^0, W_k^0, A_k) = \frac{1}{D(p^1)} \, [\ln p^{1'} \, (\alpha_p + \frac{1}{2} B_{pp} \ln p^1) - W_k^0]$$

$$(4.4)$$

$$+ \ln m_0(p^1, A_k) - \ln M_k^0, \quad (k = 1, 2 \ldots K).$$

We can refer to the index P_k as the **translog individual cost-of-living index**. If the translog
index is greater than unity and total expenditure is constant, then the welfare of the consuming unit is decreased by the change in prices.

We can illustrate the measurement of the individual cost-of-living index by representing
the impact of a change in prices in diagrammatic form. For simplicity we consider the case
of two commodities (N = 2). In Diagram 1 we have depicted the indifference map of the
kth household. Consumer equilibrium at the base price system p^0 is represented by the
point A with base level of individual welfare W_k^0. The corresponding level of total expenditure M_k^0, divided by the base price of the second commodity p_2^0, is given on the vertical
axis. This axis provides a representation of total expenditure in terms of units of the second commodity.

Consumer equilibrium after the change in prices is represented by the point B with
associated level of individual welfare W_k^1. The level of total expenditure associated with
the change in prices M_k^1, divided by the current price of the second commodity p_2^1, is given,
as before, on the vertical axis. Finally, the level of total expenditure required to attain
the base level of individual welfare W_k^0 at current prices $M(p^1, W_k^0, A_k)$ corresponds to
consumer equilibrium at the point C. The individual cost- of-living index is given by ratio
of the distances on the vertical axis corresponding to the consumer equilibrium at points
C and A, multiplied by the ratio of prices of the second commodity at the two points p_2^1 / p_2^0.

As a further illustration of the individual cost-of-living index, we analyze the changes
in prices over the period 1958-1978, using prices for 1972 as the base price system. For
this purpose we employ the econometric model of aggregate consumer behavior presented

Diagram 1.   Individual Cost of Living Index.

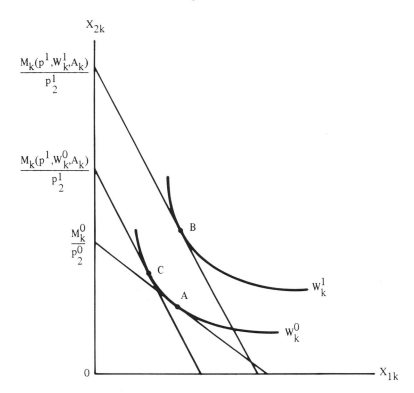

in Section 3. This model is based on time series and cross-section data on personal consumption expenditures for the United States, broken down by five commodity groups. The five commodity groups are energy, food, other consumer goods, capital services and other consumer services.

Using the translog individual cost-of-living index (4.4), we can assess the impact of price changes on households with different base levels of individual welfare and different demographic characteristics. For this purpose we set the base level of individual welfare at the levels attained in 1972 with half mean expenditure in that year of $4,467, mean expenditure or $8,934, and twice mean expenditure or $17,868. We present translog individual cost-of-living indexes for the period 1958-1978 for white and nonwhite households with urban and rural residences in Appendix Table 1. Within each of these groups we consider

TABLE 2. Changes in Individual Cost-of-Living Indexes (annual percentage rates)

Year	Urban		Rural	
	White	Nonwhite	White	Nonwhite
1959	− 0.02	− 0.05	0.24	0.21
1960	2.94	2.95	2.67	2.68
1961	0.84	0.83	0.86	0.85
1962	1.03	1.02	1.01	1.00
1963	0.46	0.46	0.62	0.62
1964	2.19	2.23	1.91	1.95
1965	3.26	3.26	2.94	2.93
1966	2.97	2.95	3.01	2.99
1967	0.73	0.74	1.09	1.10
1968	3.16	3.21	3.24	3.29
1969	6.18	6.23	5.76	5.80
1970	1.48	1.43	2.15	2.10
1971	3.78	3.76	3.77	3.75
1972	7.05	7.07	6.29	6.31
1973	8.21	8.14	8.00	7.93
1974	8.87	8.56	10.33	10.03
1975	4.65	4.60	5.40	5.34
1976	6.39	6.43	6.10	6.14
1977	8.36	8.31	7.91	7.86
1978	7.04	7.02	6.90	6.88

Region = Northeast.
Size = 5.
Age = 35-44.
Expenditure = $4467. in 1972.

TABLE 2. Changes in Individual Cost-of-Living Indexes (annual percentage rates) – Continued

Year	Urban		Rural	
	White	Nonwhite	White	Nonwhite
1959	0.08	0.06	0.35	0.33
1960	2.97	2.97	2.69	2.70
1961	0.84	0.83	0.86	0.86
1962	1.06	1.05	1.04	1.03
1963	0.49	0.49	0.65	0.65
1964	2.21	2.24	1.93	1.96
1965	3.18	3.17	2.85	2.84
1966	2.86	2.84	2.90	2.88
1967	0.86	0.87	1.23	1.24
1968	3.26	3.31	3.35	3.39
1969	6.21	6.25	5.79	5.83
1970	1.49	1.44	2.16	2.10
1971	3.91	3.89	3.90	3.88
1972	6.96	6.97	6.20	6.22
1973	7.71	7.63	7.50	7.43
1974	8.15	7.84	9.61	9.31
1975	4.68	4.63	5.43	5.37
1976	6.79	6.82	6.50	6.53
1977	8.48	8.43	8.03	7.98
1978	6.70	6.68	6.55	6.53

Region = Northeast.
Size = 5.
Age = 35-44.
Expenditure = $8,934 in 1972.

TABLE 2. Changes in Individual Cost-of-Living Indexes (annual percentage rates) – Concluded

Year	Urban		Rural	
	White	Nonwhite	White	Nonwhite
1959	0.20	0.17	0.47	0.44
1960	2.99	2.99	2.72	2.72
1961	0.84	0.84	0.87	0.86
1962	1.08	1.07	1.07	1.06
1963	0.52	0.52	0.68	0.68
1964	2.23	2.26	1.95	1.98
1965	3.09	3.08	2.76	2.75
1966	2.75	2.73	2.79	2.78
1967	1.00	1.01	1.37	1.38
1968	3.37	3.42	3.45	3.50
1969	6.24	6.28	5.81	5.85
1970	1.50	1.44	2.16	2.11
1971	4.04	4.02	4.03	4.01
1972	6.86	6.88	6.11	6.12
1973	7.21	7.13	7.00	6.92
1974	7.43	7.12	8.90	8.59
1975	4.71	4.66	5.45	5.40
1976	7.18	7.21	6.89	6.92
1977	8.59	8.55	8.15	8.10
1978	6.35	6.33	6.21	6.19

Region = Northeast.
Size = 5.
Age = 35-44.
Expenditure = $17,868 in 1972.

families of size five with a head of household aged 35-44, living in the Northeast region of the United States.

We present rates of inflation calculated from the translog individual cost-of-living indexes in Table 2. For example, the change in the logarithm of the translog index between 1958 and 1959 for a white, urban family with total expenditure equal to $4,467 in 1972 is − 0.02 percent. The corresponding change for a white, rural family is 0.24 percent. For nonwhite households the change in cost of living for urban households is − 0.05 percent, while the change for rural households is 0.21 percent. If we compare translog individual cost-of-living indexes for households with different base levels of total expenditure, we find that rates of inflation are greater for higher levels of expenditure for the periods 1958-1964, 1966-1971, and 1974-1977. For the remainder of the period -- 1964-1966, 1971-1974, and 1977-78 -- rates of inflation are greater for lower levels of expenditure.

We can compare rates of inflation for households with different demographic characteristics. For example, a white, urban family with mean base level of expenditure experienced a change of 8.15 percent in the cost-of-living between 1973 and 1974. By comparison a white, rural household at the same level of expenditure had a change of 9.61 percent over the same period. On the basis of the results presented in Table 2 we conclude that there are substantial differences in rates of inflation for households with different base levels of individual welfare and with different demographic characteristics.

To summarize: We have defined the individual cost of living index as the ratio of the total expenditure required to attain a base level of individual welfare at current prices to the base level of expenditure. Using the translog indirect utility function (2.22) and the translog expenditure function (2.23), we implement this definition by means of the translog individual cost of living index (4.4). We find that changes in the individual cost-of-living vary substantially for households with different base levels of welfare and different demographic characteristics over the period 1958-1978.

5. Social Welfare Functions

Our next objective is to generate a class of possible social welfare functions that can provide the basis for social cost of living measurement. For this purpose we must choose social welfare functions capable of expressing the implications of a variety of different ethical judgements. To facilitate comparisons with alternative approaches, we employ the axiomatic framework for social choice used by Arrow [1963], Sen [1970], and Roberts [1980a] in proving the impossibility of a nondictatorial social ordering.

We consider the set of all possible social orderings over the set of social states, say X, and the set of all possible real-valued individual welfare functions, say W_k (k = 1, 2 ... K). A social ordering, say R, is a complete, reflexive, and transitive ordering of social states. A social state is described by the quantities consumed of N commodity groups by K individuals. The individual welfare function for the kth individual W_k (k = 1, 2 ... K) is defined on the set of social states X and gives the level of individual welfare for that individual in each state.

To describe social orderings in greater detail we find it useful to introduce the following notation:

x_{nk} – the quantity of the nth commodity group consumed by the kth consuming unit (n = 1, 2 ... N; k = 1, 2 ... K).

x – a matrix with elements $\{x_{nk}\}$ describing the social state.

$u = (W_1, W_2 ... W_K)$ – a vector of individual welfare functions of all K individuals.

Following Sen [1970, 1977] and Hammond [1976] we define a **social welfare functional**, say f, as a mapping from the set of individual welfare functions to the set of social orderings, such that $f(u') = f(u)$ implies $R' = R$, where:

$$u = [W_1(x), W_2(x) ... W_K(x)],$$

$$u' = [W_1'(x), W_2'(x) ... W_K'(x)],$$

for all $x \in X$. Similarly, we define L_k ($k = 1, 2 \ldots K$) as the **set of admissible individual welfare functions** for the kth individual and L as the Cartesian product $\prod_{k=1}^{K} L_k$. Finally let $\underline{L}$ be the partition of L such that all elements of $\underline{L}$ yield the same social ordering.

We can describe a social ordering in terms of the following properties of a social welfare functional:

1. Unrestricted Domain. The social welfare functional f is defined for all possible vectors of individual welfare functions u.

2. Independence of Irrelevant Alternatives. For any subset A contained in X, if $u(x) = u'(x)$ for all $x \in A$, then $R:A = R':A$, where $R = f(u)$ and $R' = f(u')$ and $R:A$ is the social ordering over the subset A.

3. Positive Association. For any vectors of individual welfare functions u and u', if for all y in X-x, such that:

$$W'_k(y) = W_k(y),$$

$$W'_k(x) > W_k(x), \qquad (k = 1, 2 \ldots K),$$

then xPy implies $xP'y$ and $yP'x$ implies yPx, where P is a strict ordering of social states.

4. Nonimposition. For all x, y in X there exist u, u' such that xPy and $yP'x$.

5. Cardinal Full Comparability. The set of admissible individual welfare functions that yield the same social ordering $\underline{L}$ is defined by:

$$\underline{L} = \left\{ u': W'_k(x) = \alpha + \beta W_k(x), \beta > 0, k = 1, 2 \ldots K \right\},$$

and $f(u') = f(u)$ for all $u' \in \underline{L}$.

Cardinal full comparability implies that social orderings are invariant with respect to any positive affine transformation of the individual welfare functions $\{ W_k \}$ that is the same for all individuals. By contrast Arrow requires ordinal noncomparability,[15] which implies that social orderings are invariant with respect to monotone increasing transformations of the individual welfare functions that may differ among individuals:

5′. Ordinal Noncomparability. The set of individual welfare functions that yield the same social ordering $\underline{L}$ is defined by:

$$\underline{L} = \left\{ u': W'_k(x) = \phi_k [W_k(x)], \phi_k \text{ increasing, } k = 1, 2 \dots K \right\},$$

and $f(u') = f(u)$ for all u' in $\underline{L}$.

The properties of a social welfare functional corresponding to unrestricted domain and independence of irrelevant alternatives are used by Arrow in proving the impossibility of a nondictatorial social ordering:

4′. Nondictatorship. There is no individual k such that for all x, y ϵ X, $W_k(x) > W_k(y)$ implies xPy.

Under ordinal noncomparability the assumptions of positive association and nonimposition employed by Arrow imply the weak Pareto principle:

3′ Pareto Principle. For any x, y ϵ X, if $W_k(x) > W_k(y)$ for all individuals (k = 1, 2 ... K), then xPy.

If a social welfare functional f has the properties of unrestricted domain, independence of irrelevant alternatives, the weak Pareto principle, and ordinal noncomparability, then no nondictatorial social ordering is possible. This result is Arrow's impossibility theorem. Since it is obvious that the class of dictatorial social orderings is too narrow to provide an adequate basis for expressing the implications of alternative ethical judgments, we propose to generate a class of social welfare functions suitable for the evaluation of alternative economic policies by weakening Arrow's assumptions.

We first consider weakening the assumption of ordinal noncomparability of individual welfare functions. Sen [1970] has shown that Arrow's conclusion that no nondictatorial social ordering is possible is preserved by replacing ordinal noncomparability by cardinal noncomparability. This implies that social orderings are invariant with respect to positive affine transformations of the individual welfare functions that may differ among individuals:

5′′. Cardinal Noncomparability. The set of individual welfare functions that yield same social ordering $\underline{L}$ is defined by:

$$\underline{L} = \left\{ u': W'_k(x) = \alpha_k + \beta_k W_k(x), \beta_k > 0, k = 1, 2 \dots K \right\},$$

and $f(u') = f(u)$ for all u' in $\underline{L}$.

However, d'Aspremont and Gevers [1977], Deschamps and Gevers [1978], Maskin [1978] and Roberts [1980b] have shown that we obtain an interesting class of nondictatorial social orderings by requiring cardinal unit comparability of individual welfare functions, which implies that social orderings are invariant with respect to positive affine transformations with units that are the same for all individuals:

5′′′. Cardinal Unit Comparability. The set of individual welfare functions that yield the same social ordering $\underline{L}$ is defined by:

$$\underline{L} = \left\{ u': W'_k(x) = \alpha_k + \beta W_k(x), \beta > 0, k = 1, 2 \dots K \right\},$$

and $f(u') = f(u)$ for all u' in $\underline{L}$.

If a social welfare functional f has the properties of unrestricted domain, independence of irrelevant alternatives, the weak Pareto principle, and cardinal unit comparability, there exist social orderings and a continuous real-valued social welfare function, say W, such that if $W[u(x)] > W[u(y)]$, then xPy. Furthermore, the social welfare function can be represented in the form:

$$W[u(x)] = \sum_{k=1}^{K} a_k W_k(x). \tag{5.1}$$

If we add the assumption that the social welfare function has the property of anonymity, that is, no individual is given greater weight than any other individual in determining the level of social welfare, then the social welfare function W in (5.1) must be symmetric in the individual welfare functions $\{W_k\}$. The property of anonymity incorporates a notion of horizontal equity into the representation of social orderings.

Under anonymity the function W in (5.1) reduces to the sum of individual welfare functions and takes the form of a utilitarian social welfare function. Utilitarian social welfare functions have been employed extensively in applications of welfare economics, especially in the measurement of inequality by methods originated by Atkinson [1970] and Kolm [1969, 1976a, 1976b], in the design of optimal income tax schedules along the lines pioneered by Mirrlees [1971], and in the evaluation of alternative economic policies by Arrow and Kalt [1979].

The approach to the measurement of social welfare based on a utilitarian social welfare function provides a worthwhile starting point for applications. Harsanyi [1976] and Ng [1975] have pointed out that distributional considerations can be incorporated into a utilitarian social welfare function through the representation of individual welfare functions. However, Sen [1973, p.18] has argued that a utilitarian social welfare function does not take appropriate account of the distribution of welfare among individuals:

> The distribution of welfare between persons is a relevant aspect of any problem of income distribution, and our evaluation of inequality will obviously depend on whether we are concerned only with the loss of the sum of individual utilities through a bad distribution of income, or also with the inequality of welfare levels of different individuals.

To broaden the range of possible social orderings we can require cardinal full comparability of individual welfare functions, as defined above. Roberts [1980b] has shown that a social welfare functional f with the properties of unrestricted domain, independence of irrelevant alternatives, the weak Pareto principle, and cardinal full comparability implies the existence of a social welfare function that takes the form:

$$W[u(x)] = \bar{W}(x) + g[u(x) - \bar{W}(x) \, i],\qquad (5.2)$$

where i is a vector of ones, the function $\bar{W}(x)$ corresponds to average individual welfare:

$$\bar{W}(x) = \sum_{k=1}^{K} a_k W_k(x),$$

and $g(x)$ is a linear homogeneous function of deviations of levels of individual welfare from the average.[16]

If the function $g(x)$ in the representation (5.2) of the social welfare function is identically equal to zero, then the social welfare function reduces to the form (5.1). If the function $g(x)$ is not identically zero, then the social welfare function incorporates both a measure of average individual welfare and a measure of inequality in the distribution of individual welfare. We conclude that the class of possible social welfare functions (5.2) includes utilitarian welfare functions, but also includes functions that are not subject to the objections that can be made to utilitarianism.

Although Roberts [1980b] has succeeded in broadening the class of possible social welfare functions beyond those consistent with utilitarianism, the social welfare functions (5.2) are subject to an objection raised by Sen [1973].[17] Information about alternative social states enters only through the individual welfare functions $\{W_k\}$. Sen refers to this property of a social welfare function f as **welfarism**. Welfarism rules out characteristics of a social state that are conceivably relevant for social orderings, but that cannot be incorporated into the social welfare function through the individual welfare functions.

Roberts [1980b] has suggested the possibility of further weakening of Arrow's assumptions in order to incorporate nonwelfare characteristics of social states.[18] For this purpose we can replace the weak Pareto principle by positive association and nonimposition, as defined above. We retain the assumptions of unrestricted domain, independence of irrelevant alternatives, and cardinal full comparability of measures of individual welfare. We can partition the set of social states X into subsets, such that all states within each subset have the same nonwelfare characteristics. For each subset there exists a social ordering that can be represented by a social welfare function of the form (5.2).

Under the assumptions we have outlined there exists a social ordering for the set of all social states that can be represented by a social welfare function of the form:

$$W(u, x) = F\left\{\bar{W}(x) + g[x, u(x) - \bar{W}(x) i], x\right\}, \qquad (5.3)$$

where the function $\bar{W}(x)$ corresponds to average individual welfare:

$$\bar{W}(x) = \sum_{k=1}^{K} a_k(x) W_k(x).$$

As before, the function g is a linear homogeneous function of deviations of levels of individual welfare from average welfare.

The class of social welfare functions (5.3) incorporates nonwelfare characteristics of social states through the weights $\{a_k(x)\}$ in average individual welfare $\bar{W}(x)$, through the function g(x), which depends directly on the social state x as well as on deviations of levels of individual welfare from the average welfare, and through the function F, which depends directly on the social state x and on the sum of the functions $\bar{W}(x)$ and g(x). This class includes social welfare functions that are not subject to the objections that can be made to welfarism.

At this point we have generated a class of possible social welfare functions capable of expressing the implications of a variety of different ethical judgments. In order to choose a specific social welfare function, we must narrow the range of possible ethical judgments by imposing further requirements on the class of possible social welfare functions. First, we must limit the dependence of the function F(x) in (5.3) on the characteristics of alternative social states. Second, we must select a form for the function g(x) in (5.3), which depends on deviations of levels of individual welfare from average welfare $\bar{W}(x)$. Finally, we must choose representations of the individual welfare functions $\{W_k(x)\}$ that provide cardinal full comparability.

We first rule out the dependence of the function F(x) in (5.3) on characteristics of social states that do not enter through the functions $\bar{W}(x)$ and g(x). This restriction reduces F to a function of a single variable $\bar{W} + g$. We obtain an ordinal measure of social welfare by permitting the function F to be any monotone increasing transformation. To obtain a cardinal measure of social welfare we observe that the function $\bar{W}(x) + g$ is homogeneous of degree one in the individual welfare functions $\{W_k(x)\}$. All representations of the social welfare function that preserve this property can be written in the form:

$$W(u, x) = \beta[\bar{W}(x) + g(x)], \beta > 0. \tag{5.4}$$

We conclude that only positive, homogeneous, affine transformations are permitted.

The restrictions embodied in the class of social welfare functions (5.4) do not reduce social welfare to a function of the individual welfare functions $\{W_k(x)\}$ alone, since the weights $\{a_k(x)\}$ in average individual welfare $\bar{W}(x)$ and the function g(x) depend on nonwelfare characteristics of the social state x. However, these social welfare functions are homogeneous of degree one in levels of individual welfare. This implies that doubling the welfare of each individual will double social welfare, holding nonwelfare characteristics of the social state constant. Blackorby and Donaldson [1982] refer to this class of social welfare functions as **distributionally homothetic**.[19]

We impose a second set of requirements on the class of social welfare functions (5.3) by selecting an appropriate form for the function g(x). In particular, we require that this function is additive in deviations of individual welfare functions $\{W_k(x)\}$ from average welfare $\bar{W}(x)$. Since the function g(x) is homogeneous of degree one, it must be a mean value function of order $\rho(x)$:[20]

$$g[x, u(x) - \bar{W}(x) i] = -\gamma(x)[\sum_{k=1}^{K} b_k(x) |W_k - \bar{W}|^{-\rho(x)}]^{-\frac{1}{\rho(x)}}, \qquad (5.5)$$

where:

$$\gamma(x) > 0, \rho(x) \le -1, \sum_{k=1}^{K} b_k(x) = 1, 0 < b_k(x) < 1, (k = 1, 2 \dots K).$$

Under these restrictions the function $g(x)$ is negative, except at the point of perfect equality $W_k = \bar{W}$ $(k = 1, 2 \dots K)$, where it is zero.

The function (x) in the representation (5.5) determines the curvature of the social welfare function in the individual welfare functions $\{W_k(x)\}$. We can refer to this function as the **degree of aversion to inequality**. We assume that this function is constant, so that the corresponding social welfare function $W(u,x)$ is characterized by a constant degree of aversion to inequality. To complete the selection of an appropriate form for the social welfare function we must choose appropriate weights $\{a_k(x)\}$ for average individual welfare $\bar{W}(x)$ and $\{b_k(x)\}$ for the measure of equality $g(x)$. We find it natural to require that the two sets of weights are the same.

To incorporate a notion of horizontal equity into the social welfare functions (5.5) we can impose a weak form of the property of anonymity. In particular, we require that no individual is given greater weight in the social welfare function than any other individual with an identical individual welfare function. This implies that the social welfare function is symmetric in the levels of individual welfare for identical individuals. The weights $\{a_k(x)\}$ in average welfare $\bar{W}(x)$ and the measure of equality $g(x)$ must be the same for identical individuals.

Under the restrictions presented up to this point the social welfare function W takes the form:

$$W(u,x) = \bar{W} - \gamma(x) \left[\sum_{k=1}^{K} a_k(x) |W_k - \bar{W}|^{-\rho} \right]^{-\frac{1}{\rho}} \qquad (5.6)$$

where:

$$\bar{W}(x) = \sum_{k=1}^{K} a_k(x) W_k(x).$$

The condition of positive association requires that an increase in all levels of individual welfare must increase social welfare. This condition implies that the average level of individual welfare $\bar{W}$ must increase by more than the function $g(x)$, whatever the initial distribution of individual welfare. We assume that the function $\gamma(x)$ in (5.6) must take the maximum value consistent with positive association, so that:

$$\gamma(x) = \left\{ 1 + \left[\frac{\sum_{k=1}^{K} a_k(x)}{a_j(x)} \right] - (\rho + 1) \right\}^{\frac{1}{\rho}}, \qquad (5.7)$$

where:

$$a_j(x) = \min_k a_k(x), \qquad (k = 1, 2 \ldots K),$$

for the social state x.

To complete the selection of a social welfare function $W(u,x)$ we require that the individual welfare functions $\{W_k\}$ in (5.3) must be invariant with respect to any positive affine transformation that is the same for all households.[21] Under this assumption the logarithm of the translog indirect utility function is a cardinal measure of individual welfare with full comparability among households. The social welfare function takes the form:

$$W(u,x) = \ln \bar{V} - \gamma(x) \left[\sum_{k=1}^{K} a_k(x) |\ln V_k - \ln \bar{V}|^{-\rho} \right]^{-\frac{1}{\rho}}. \tag{5.8}$$

where:

$$\ln \bar{V} = \sum_{k=1}^{K} a_k(x) \ln V_k \left[\frac{p\, m(A_k)}{M_k} \right].$$

We can complete the specification of a social welfare function $W(u,x)$ by choosing a set of weights $a_k(x)$ for the levels of individual welfare $\left\{ \ln V_k \left[\dfrac{p\, m(A_k)}{M_k} \right] \right\}$ in (5.8). For this purpose we must appeal to a notion of vertical equity. Following Hammond [1977], we define a distribution of total expenditure $\{M_k\}$ as more **equitable** than another distribution $\{M'_k\}$ if:

(i) $M_i + M_j = M'_i + M'_j$,

(ii) $M_k = M'_k$ for $k \neq i, j$,

(iii) $\ln V_i \left[\dfrac{p\, m(A_i)}{M'_i} \right] > \ln V_i \left[\dfrac{p\, m(A_i)}{M_i} \right] > \ln V_i \left[\dfrac{p\, m(A_j)}{M_j} \right] > \ln V_j \left[\dfrac{p\, m(A_j)}{M'_j} \right]$,

We say that a social welfare function $W(u,x)$ is **equity-regarding** if it is larger for a more equitable distribution of total expenditure.

We require that the social welfare function (5.8) must be equity-regarding. This amounts to imposing a version of Dalton's [1920] principle of transfers. This principle requires that a transfer of total expenditures from a rich household to a poor household that does not reverse their relative positions in the distribution of total expenditure must increase the level of social welfare.

If the social welfare function (5.8) is required to be equity-regarding, then the weights $\{a_k(x)\}$ associated with the individual welfare functions $\left\{\ln V_k\left[\dfrac{p\,m(A_k)}{M_k}\right]\right\}$ must take the form:

$$a_k(x) = \frac{m_0(p,A_k)}{\displaystyle\sum_{k=1}^{K} m_0(p,A_k)}\ , \qquad (k = 1, 2 \ldots K). \tag{5.9}$$

We conclude that an equity-regarding social welfare function of the class (5.8) must take the form:

$$W(u,x) = \ln \bar{V} - \gamma(x)\left[\frac{\displaystyle\sum_{k=1}^{K} m_0(p,A_k)|\ln V_k - \ln \bar{V}|^{-\rho}}{\displaystyle\sum_{k=1}^{K} m_0(p,A_k)}\right]^{-\frac{1}{\rho}}, \tag{5.10}$$

where:

$$\ln \bar{V} = \frac{\displaystyle\sum_{k=1}^{K} m_0(p,A_k)\,\ln V_k\left[\dfrac{p\,m(A_k)}{M_k}\right]}{\displaystyle\sum_{k=1}^{K} m_0(p,A_k)}\ ,$$

$$= \ln p'\,(\alpha_p + \frac{1}{2}B_{pp}\ln p) - D(p)\,\frac{\displaystyle\sum_{k=1}^{K} m_0(p,A_k)\,\ln[M_k/m_0(p,A_k)]}{\displaystyle\sum_{k=1}^{K} m_0(p,A_k)}\ .$$

Furthermore, the condition of positive association implies that the function $\gamma(x)$ in (5.10) must take the form:

$$
\gamma(x) = \left\{ 1 + \left[\frac{\displaystyle\sum_{k=1}^{K} m_0(p,A_k)}{m_0(p,A_j)} \right]^{-(\rho+1)} \right\}^{\frac{1}{\rho}} , \qquad (5.11)
$$

where:

$$
m_0(p,A_j) = \min_k m_0(p,A_k), \quad (k = 1, 2 \dots K).
$$

In order to formulate a social cost-of-living index, we can introduce the social expenditure function, defined as the minimum level of aggregate expenditure $M = \sum_{k=1}^{K} M_k$ required to attain a given level of social welfare, say W, at a specified price system p.[22] More formally, the social expenditure function $M(p, W)$ is defined by:

$$
M(p,W) = \min \left\{ M : W(u,x) \geq W; \; M = \sum_{k=1}^{K} M_k \right\}. \qquad (5.12)
$$

The social expenditure function (5.12) is precisely analogous to the individual expenditure function (2.19). The individual expenditure function gives the minimum level of individual expenditure required to attain a stipulated level of individual welfare; the social expenditure function gives the minimum level of aggregate expenditure required to attain a stipulated level of social welfare. Just as the individual expenditure function and the indirect utility function can be employed in measuring the individual cost of living, the social expenditure function and the social welfare function can be employed in measuring the social cost of living.

To construct a social expenditure function we first maximize social welfare for a fixed level of aggregate expenditure. We can maximize the average level of individual welfare for a given level of aggregate expenditure by means of the Lagrangian:

$$Z = \ln \bar{V} + \lambda [\sum_{k=1}^{K} M_k - M],$$ (5.13)

$$= \ln p'(\alpha_p + \frac{1}{2} B_{pp} \ln p) - D(p) \frac{\sum_{k=1}^{K} m_0(p,A_k) \ln [M_k/m_0(p,A_k)]}{\sum_{k=1}^{K} m_0(p,A_k)}$$

$$+ \lambda [\sum_{k=1}^{K} M_k - M].$$

The first-order conditions for a constrained maximum of average individual welfare are:

$$\frac{D(p)}{\sum_{k=1}^{K} m_0(p,A_k)} \cdot \frac{m_0(p,A_k)}{M_k} = \lambda ,$$

$$\sum_{k=1}^{K} M_k = M, \qquad (k = 1, 2 \ldots K),$$

so that total expenditure per household equivalent member $\{ M_k/m_0(p,A_k)\}$ is the same for all consuming units.

Next we consider the class of social welfare functions (5.8). Since the function $g(x, u - \bar{W} \ i)$ is nonpositive, we obtain a maximum of the social welfare function (5.8) if the function $g(x, u - \bar{W} \ i)$ can be made equal to zero while the average level of individual welfare $\bar{W}$ is a maximum. If total expenditure per household equivalent member $\{ M_k/m_0(p,A_k)\}$ is the same for all consuming units, the function $g(x, u - \bar{W} \ i)$ is equal to zero, so that the social welfare function $W(u,x)$ in (5.8) is a maximum.

If aggregate expenditure is distributed so as to equalize total expenditure per household equivalent member, the level of individual welfare is the same for all consuming units. For this distribution of total expenditure the social welfare function (5.8) reduces to the average level of individual welfare $\ln \bar{V}$. For the translog indirect utility function the maximum value of social welfare for a given level of aggregate expenditure takes the form:

$$W(x,u) = \ln \bar{V}.$$ (5.14)

$$= \ln p'(\alpha_p + \frac{1}{2} B_{pp} \ln p) - D(p) \ln [M/ \sum_{k=1}^{K} m_0(p,A_k)].$$

This value of social welfare is obtained by evaluating the translog indirect utility function (2.22) at total expenditure per household equivalent member $M/ \sum_{k=1}^{K} m_0(p,A_k)$ for the economy as a whole. We can solve for aggregate expenditure as a function of the level of social welfare and prices:

$$\ln M(p, W) = \frac{1}{D(p)} [\ln p'(\alpha_p + \frac{1}{2} B_{pp} \ln p) - W] + \ln [\sum_{k=1}^{K} m_0(p,A_k)].$$ (5.15)

We can refer to this function as the **translog social expenditure function**. The value of aggregate expenditure is obtained by evaluating the translog individual expenditure function (2.23) at the level of social welfare W and the number of household equivalent members $\sum_{k=1}^{K} m_0(p,A_k)$ for the economy as a whole.

To summarize: We have generated a class of social welfare functions (5.3) that has the properties of unrestricted domain, independence of irrelevant alternatives, positive association, nonimposition and cardinal full comparability. By imposing the additional assumption that the degree of aversion to inequality is constant and requiring the social welfare function to satisfy requirements of horizontal and vertical equity, we obtain the social welfare function (5.10). Finally, we have derived a social expenditure function (5.15) giving the minimum aggregate expenditure required to attain a base level of social welfare.

6. Social Cost-of-Living Index.

In this section we present methods for translating changes in prices into measures of change in the social cost of living. Following Pollak [1981], we define the social cost-of-living index, say P, as the ratio of two levels of aggregate expenditure:

$$P(p^1, p^0, w^0) = \frac{M(p^1, w^0)}{M(p^0, w^0)}. \tag{6.1}$$

In this ratio the numerator is the aggregate expenditure required to attain the base level of social welfare W^0 at the current price system p^1; the denominator is the expenditure required to attain this level of social welfare at the base price system p^0.

The social welfare function (5.10) and the translog social expenditure function (5.15) can be employed in implementing the social cost-of-living index (6.1). First, we can express the base level of social welfare W^0 in terms of the social welfare functions:

$$W^0 = \ln \bar{V}^0 - \tag{6.2}$$

$$\left\{ 1 + \left[\frac{\sum\limits_{k=1}^{K} m_0(p^0, A_k)}{m_0(p^0, A_j)} \right]^{-1} (\rho - 1) \right\}^{\frac{1}{\rho}} \left[\frac{\sum\limits_{k=1}^{K} m_0(p^0, A_k) |\ln V_k^0 - \ln \bar{V}^0|^{-\rho} - 1}{\sum\limits_{k=1}^{K} m_0(p^0, A_k)} \right]^{\frac{1}{\rho}},$$

where:

$$\ln \bar{V}^0 = \frac{\sum\limits_{k=1}^{K} m_0(p^0, A_k) \ln V_k \left[\frac{p^0 m(A_k)}{M_k^0} \right]}{\sum\limits_{k=1}^{K} m_0(p^0, A_k)},$$

and:

$$\ln V_k^0 = \ln p^{0'} (\alpha_p + \frac{1}{2} B_{pp} \ln p^0) - D(p^0) \ln [M_k^0/m_0(p^0, A_k)], \ (k = 1, 2 \ldots K).$$

Using the translog social expenditure function (5.21), we can express the social cost-of-living index (6.1) in the form:

$$\ln P(p^1, p^0, W^0) = \frac{1}{D(p^1)} [\ln p^{1'} (\alpha_p + \frac{1}{2} B_{pp} \ln p^1) - W^0] \qquad (6.3)$$

$$+ \ln [\sum_{k=1}^{K} m_0(p^1, A_k)] - \ln M^0.$$

We can refer to the index P as the **translog social cost-of-living index**. If the translog index is greater than unity and aggregate expenditure is constant, then social welfare is decreased by the change in prices.

The translog social expenditure function (5.15) has the same form as the translog individual expenditure function (2.23). We can express the level of social welfare as a function of the price system p and aggregate expenditure M. We can take this level of social welfare to be the average level of individual welfare $\bar{W}$, obtained by redistributing aggregate expenditure among households so as to equalize total expenditure per household equivalent member. Under this assumption society behaves in the same way as an individual maximizing a utility function, as demonstrated by Samuelson [1956] and Pollak [1981]. We can represent social welfare in terms of the indifference map for a single representative consumer, as in Diagram 1.

In Diagram 2 we have depicted the indifference map of the representative consumer with indirect utility function given by the average level of utility $\ln \bar{V}$ in (5.14). As before, we consider the case of two commodities ($N = 2$) for simplicity. Equilibrium of the representative consumer at the base price system p^0 is represented by the point A with base level of average welfare $\bar{W}^0$. The corresponding level of aggregate expenditure M^0, divided by the base price of the second commodity p_2^0, is given on the vertical axis. This axis provides a representation of aggregate expenditure in terms of units of the second commodity.

Diagram 2.   Social Cost of Living Index.

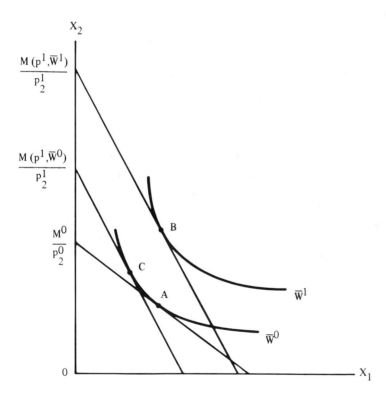

Equilibrium of the representative consumer after the change in prices is represented by the point B with associated level of average welfare $\overline{W}^1$. The level of aggregate expenditure associated with the change in prices M^1, divided by the current price of the second commodity p_2^1, is given, as before, on the vertical axis. Finally, the level of aggregate expenditure required to attain the base level of average welfare $\overline{W}^0$ at current prices $M(p^1, \overline{W}^0)$, corresponds to equilibrium of the representative consumer at the point C. The social cost-of-living index is given by the ratio of the distances on the vertical axis corresponding to consumer equilibrium at points C and A, multiplied by the ratio of prices of the second commodity at the two points p_2^1 / p_2^0.

As a further illustration of the social cost-of-living index, we analyze the changes in prices over the period 1958-1978, using prices for 1972 as the base price system. As before, we employ the econometric model of aggregate consumer behavior presented in Section 3 for this purpose. Using the translog social cost-of-living index (6.3), we can assess the impact of price changes on the U.S. economy as a whole. We present the social cost-of-living index and rates of inflation corresponding to this index in Table 3. Over the 20-year period 1958-1978 this index has risen from .6928 to 1.5214 with 1972 equal to 1.0000.

To summarize: We have defined the social cost-of-living index as the ratio of the aggregate expenditure required to attain a base level of social welfare at current prices to the base level of expenditure. Using the average level of social welfare (5.14) and the translog social expenditure function (5.15), we implement this definition by means of the translog social cost-of-living index (6.3). We illustrate the translog index by analyzing changes in prices over the period 1958-1978 for the U.S. economy as a whole.

TABLE 3. Social Cost-of-Living Index

Year	Social Cost-of-Living Index (1972 = 1.0000)	Inflation Rate (annual percentage rates)
1958	0.6928	0.00
1959	0.6946	0.25
1960	0.7156	2.96
1961	0.7217	0.85
1962	0.7296	1.09
1963	0.7335	0.53
1964	0.7497	2.17
1965	0.7731	3.07
1966	0.7947	2.75
1967	0.8030	1.03
1968	0.8303	3.34
1969	0.8831	6.16
1970	0.8972	1.58
1971	0.9344	4.06
1972	1.0000	6.77
1973	1.0748	7.21
1974	1.1619	7.79
1975	1.2194	4.83
1976	1.3101	7.17
1977	1.4281	8.61
1978	1.5214	6.33

7. Group Cost-of-Living Indexes

In Sections 4 and 6 we have presented cost-of-living indexes for individual households and for the U.S. economy as a whole. In this section our objective is to provide measures of the cost of living for groups of households. For this purpose we introduce group welfare and expenditure functions that are precisely analogous to the social welfare and expenditure functions of Section 5. We consider a group of G individuals, where $1 \le G \le K$;

without loss of generality we can take the group to be comprised of the first G individuals in society.

To describe group orderings we find it useful to introduce the following notation: x_{ng} – the quantity of the nth commodity group consumed by the gth consuming unit (n = 1, 2 ... N; g = 1, 2 ... G).

x_G – a matrix with elements $\{x_{ng}\}$ describing the group state.

$u_G = (W_1, W_2 ... W_G)$ – a vector of individual welfare functions for all G individuals.

We can define a group welfare functional as a mapping from the set of individual welfare functions to the set of group orderings. We can describe a group ordering in terms of properties of a group welfare functional that are precisely analogous to the properties of a social welfare functional considered in Section 5: unrestricted domain, independence of irrelevant alternatives, positive association, nonimposition and cardinal full comparability. Under these assumptions there exists a group ordering for the set of all group states that can be represented by a group welfare function analogous to the social welfare function (5.3):

$$W_G(u_G, x_G) = F\left\{\bar{W}_G(x_G) + g[x_G, u_G(x_G) - \bar{W}_G(x_G) \text{ i}], x_G\right\},\qquad (7.1)$$

where the function $\bar{W}_G(x_G)$ corresponds to average individual welfare:

$$\bar{W}_G(x_G) = \sum_{g=1}^{G} a_g(x_G) W_g(x_G).$$

As before, the function g is a linear homogeneous function of deviations of levels of individual welfare from average welfare for the group.

We can rule out the direct dependence of the function $F(x_G)$ in (7.1) on characteristics of groups states x_G. As in Section 5, this results in a cardinal representation of group welfare. To complete the selection of a group welfare function $W_G(u_G, x_G)$ we require, as before, that the individual welfare functions $\{W_k(x_G)\}$ must be invariant with respect to any positive affine transformation that is the same for all households. Second, we require that the group welfare function is equity-regarding. Under these assumptions the group welfare function must take forms analogous to (5.10).

In order to formulate a group cost-of-living index, we can introduce a group expenditure function, defined as the minimum level of group expenditure $M_G = \sum_{g=1}^{G} M_g$ required to attain a given level of group welfare, say W_G, at a specified price system p. We can maximize group welfare for a fixed level of group expenditure by equalizing total expenditure per household equivalent member. For the translog indirect utility function the group welfare function takes the form:

$$W_G(u_G, x_G) = \ln \bar{V}_G ,\qquad\qquad(7.2)$$

$$= \ln p' (\alpha_p + \tfrac{1}{2} B_{pp} \ln p) - D(p) \ln[M_G / \sum_{g=1}^{G} m_0(p, A_g),$$

where $\ln \bar{V}_G$ is the average level of individual welfare.

We can solve for group expenditure as a function of the level of group welfare and prices:

$$\ln M_G(p, W_G) = \frac{1}{D(p)} [\ln p' (\alpha_p + \tfrac{1}{2} B_{pp} \ln p) - W_G] + \ln [\sum_{g=1}^{G} m_0(p, A_g)]. \quad(7.3)$$

We can refer to this function as the **translog group expenditure function**. The value of group expenditure is obtained by evaluating the translog individual expenditure function (2.23) at the level of group welfare W_G and the number of household equivalent members $\sum_{g=1}^{G} m_0(p, A_g)$ for the group as a whole.

We can define the group cost-of-living index, say P_G, as the ratio of two levels of group expenditure:

$$P_G(p^1, p^0, W_G^0) = \frac{M_G(p^1, W_G^0)}{M_G(p^0, W_G^0)}.$$ (7.4)

In this ratio the numerator is the group expenditure required to attain the base level of group welfare W_G^0 at the current price system p^1; the denominator is the expenditure required for this level of welfare at the base price system p^0. Using the translog group expenditure function (7.3), we can express the group cost-of-living index (7.4) in the form:

$$\ln P_G(p^1, p^0, W_G^0) = \frac{1}{D(p^1)}[\ln p^{1\prime}(\alpha_p + \frac{1}{2}B_{pp}\ln p^1) - W_G^0]$$ (7.5)

$$+ \ln[\sum_{g=1}^{G} m_0(p^1, A_g)] - \ln M_G^0.$$

We can refer to the index P_G as the **translog group cost-of-living index**. If the translog index is greater than unity and group expenditure is constant, then group welfare is decreased by the change in prices.

The translog group expenditure function (7.3) has the same form as the translog social expenditure function (5.15). To obtain the group expenditure function from the social expenditure function, we replace social welfare W and the number of household equivalent members for society as a whole $\sum_{k=1}^{K} m_0(p, A_k)$ by group welfare W_G and the number of household equivalent members for the group $\sum_{g=1}^{G} m_0(p, A_g)$.

We can express the level of group welfare as a function of the price system p and group expenditure M_G. We can take this level of welfare to be the average level of individual welfare $\bar{W}_G$, obtained by equalizing total expenditure per household equivalent member among all households in the group. Under this assumption the group behaves like an individual maximizing a utility function. We can represent group welfare in terms of the indifference map for a single representative consumer, just as we represented social welfare function in terms of such an indifference map in Diagram 2.

We can illustrate the group cost-of-living index by analyzing the changes in prices over the period 1958-1978, using prices for 1972 as a base price system. We have calculated cost-of-living indexes for 21 different groups of households, classified by demographic characteristics. These demographic characteristics include size of household, age of head of household, region of residence, race and urban versus rural residence. For this purpose we set the base level of group welfare at the level the group attained in 1972. We present translog group cost-of-living indexes for the period 1958-1978 in Appendix Table 2.

We present rates of inflation calculated from the translog group cost-of-living indexes for seven family size groups in Table 4. In virtually every year the rates of inflation increase or decrease monotonically with family size. The most dramatic changes occur between unattached individuals, families of size one, and families of size two or more. For example, in 1975 unattached individuals had an inflation rate of only 3.65 percent, while households of size two experienced an inflation rate of 4.78 percent and households with seven or more members had an inflation rate of 5.31 percent. At the opposite extreme, individuals experienced an increase in cost of living of 7.98 percent in 1972, while households of size two had a rate of 6.82 percent and households of size seven or more had only 6.33 percent.

We present the impact of changes in prices on measures of the group cost of living for age groups in Table 4. We observe that rates of inflation increase or decrease monotonically with age up to the age group 55-64. In fact, the pattern of inflation rates for different age groups is similar to that for different family sizes. The reason for this is that there is a sizable correlation between family size and age of the head of household. Younger and older age groups are associated with smaller family sizes, while families in the middle of the age range are associated with larger sizes. Finally, we present rates of inflation for groups living in different regions, the two racial groups, and urban versus rural residents in Table 4. Regional differences and differences between urban and rural residents are substantial, while racial differences are not.

To summarize: We have generated group welfare functions that are analogous to the social welfare functions (5.10). Second, we have derived a group expenditure function (7.3), giving the minimum group expenditure required to attain a base level of group welfare. Finally, we have defined group cost-of-living indexes in terms of the ratio of the group expenditure required to attain a base level of group welfare at current prices to the base level of expenditure. We illustrate group cost-of- living indexes by comparing rates of inflation among 21 demographic groups for the period 1958-1978.

TABLE 4. Change in Group Cost-of-Living Indexes (annual percentage rates)

	Family Size						
Year	1	2	3	4	5	6	7 +
1959	− 0.16	0.26	0.31	0.32	0.34	0.33	0.34
1960	3.43	3.00	2.94	2.90	2.82	2.82	2.73
1961	0.79	0.85	0.85	0.85	0.87	0.87	0.88
1962	1.09	1.09	1.09	1.09	1.08	1.08	1.08
1963	0.21	0.51	0.55	0.57	0.62	0.62	0.67
1964	2.54	2.18	2.14	2.12	2.08	2.07	2.03
1965	3.63	3.09	3.02	2.98	2.92	2.93	2.87
1966	2.63	2.72	2.73	2.76	2.82	2.84	2.93
1967	0.40	1.02	1.10	1.13	1.18	1.17	1.22
1968	3.06	3.32	3.36	3.39	3.42	3.40	3.45
1969	6.76	6.19	6.12	6.08	6.00	6.00	5.92
1970	0.38	1.50	1.64	1.73	1.93	1.93	2.17
1971	4.07	4.09	4.09	4.07	4.03	4.01	3.96
1972	7.98	6.82	6.67	6.60	6.46	6.47	6.33
1973	7.68	7.15	7.08	7.10	7.18	7.25	7.36
1974	6.23	7.68	7.84	7.95	8.28	8.38	8.69
1975	3.65	4.78	4.91	4.99	5.14	5.14	5.31
1976	7.71	7.30	7.24	7.13	6.92	6.86	6.57
1977	9.50	8.72	8.60	8.49	8.31	8.29	8.05
1978	6.58	6.27	6.24	6.26	6.34	6.38	6.50

TABLE 4. Change in Group Cost-of-Living Indexes (annual percentage rates) - Continued

	Age of Head					
Year	16-24	25-34	35-44	45-54	55-64	65 +
1959	− 0.27	0.02	0.30	0.37	0.39	0.21
1960	3.62	3.20	2.92	2.83	2.83	3.00
1961	0.76	0.81	0.85	0.87	0.87	0.85
1962	1.07	1.06	1.08	1.10	1.10	1.09
1963	0.07	0.36	0.57	0.62	0.63	0.50
1964	2.63	2.36	2.16	2.07	2.05	2.19
1965	3.82	3.34	2.99	2.92	2.91	3.13
1966	2.50	2.66	2.74	2.82	2.81	2.77
1967	0.21	0.73	1.13	1.19	1.21	0.96
1968	2.93	3.21	3.42	3.40	3.40	3.28
1969	6.94	6.46	6.13	6.00	5.97	6.20
1970	− 0.12	0.91	1.68	1.95	1.97	1.51
1971	4.10	4.03	4.06	4.06	4.07	4.06
1972	8.35	7.36	6.64	6.45	6.42	6.89
1973	7.71	7.41	7.05	7.15	7.12	7.36
1974	5.81	7.00	7.66	8.30	8.40	7.86
1975	3.25	4.21	4.92	5.16	5.20	4.74
1976	8.08	7.47	7.18	6.97	6.98	7.14
1977	9.90	9.01	8.49	8.37	8.38	8.69
1978	6.55	6.44	6.25	6.30	6.27	6.41

TABLE 4. Change in Group Cost-of-Living Indexes (annual percentage rates) - Concluded

Year	Region				Race		Residence	
	NE	NC	S	W	White	Non-White	Urban	Rural
1959	0.13	0.29	0.40	0.12	0.26	0.19	0.23	0.62
1960	3.06	2.95	2.81	3.14	2.97	2.92	2.99	2.63
1961	0.84	0.85	0.87	0.83	0.85	0.84	0.85	0.88
1962	1.08	1.09	1.08	1.10	1.09	1.06	1.09	1.08
1963	0.46	0.54	0.63	0.43	0.53	0.54	0.51	0.75
1964	2.26	2.14	2.04	2.32	2.17	2.17	2.19	1.85
1965	3.21	3.03	2.88	3.25	3.07	3.07	3.09	2.63
1966	2.76	2.74	2.79	2.70	2.75	2.81	2.75	2.78
1967	0.87	1.08	1.24	0.83	1.04	1.01	1.00	1.53
1968	3.27	3.35	3.42	3.28	3.34	3.35	3.33	3.50
1969	6.30	6.12	5.95	6.40	6.17	6.14	6.19	5.67
1970	1.32	1.62	1.98	1.17	1.57	1.62	1.52	2.42
1971	4.03	4.08	4.05	4.10	4.07	3.96	4.06	4.09
1972	7.06	6.69	6.36	7.20	6.77	6.75	6.83	5.82
1973	7.42	7.14	7.07	7.26	7.19	7.41	7.24	6.78
1974	7.53	7.90	8.40	6.93	7.77	8.05	7.70	9.23
1975	4.56	4.90	5.23	4.40	4.82	4.85	4.77	5.74
1976	7.19	7.21	6.98	7.44	7.20	6.92	7.19	6.93
1977	8.76	8.60	8.32	8.94	8.63	8.42	8.65	8.08
1978	6.46	6.27	6.25	6.35	6.31	6.50	6.35	6.04

8. Summary and Conclusion

In this paper we have presented an econometric approach to individual, group and social cost-of-living measurement. The key to our approach is provided by the translog indirect utility function (2.22) and the translog individual expenditure function (2.23). In Section 2 we show how the translog indirect utility and expenditure functions can be determined from a system of individual expenditure shares (2.5) that is integrable. We also demonstrate how the individual expenditure shares (2.5) can be recovered uniquely from the system of aggregate expenditure shares (2.25).

In Section 3 we fit an econometric model of aggregate consumer behavior that incorporates the restrictions implied by integrability of the individual expenditure shares. In

Section 4 we define a cost-of-living index for the individual consuming unit. We implement this index for the period 1958-1978, using translog indirect utility and expenditure functions for all consuming units.

We define and implement a cost-of-living index for the U.S. economy as a whole for the period 1958-1978 in Section 6. This definition is precisely analogous to the definition of an individual cost-of-living index. The role of the indirect utility function is played by an explicit social welfare function introduced in Section 5. The role of the individual expenditure function is played by the translog social expenditure function (5.15). This expenditure function is simply the translog individual expenditure function, evaluated at aggregate expenditure per household equivalent member.

In Section 7 we extend the concept of a social cost-of-living index to groups of individuals. We obtain a translog group cost-of-living index that can be expressed in terms of group welfare and expenditure functions. The group expenditure function is the translog individual expenditure function, evaluated at group expenditure per household equivalent member. We present cost-of-living indexes for 21 demographic groups for the period 1958-1978.

Next, we find it useful to compare the information required for implementation of the econometric approach to cost-of-living measurement with that required for conventional cost-of-living index numbers.[23] Our cross-section data set, the Survey of Consumer Expenditures for 1972, was assembled for the purpose of providing weights for the Consumer Price Index in the United States. We have employed this data set in estimating the impact of total expenditures and demographic characteristics of households on individual expenditure patterns.

Conventional cost-of-living index numbers are based on weighted averages of price relatives for individual commodity groups. The weights are based on averages of expenditure shares for groups of consumers. Measures of the cost-of-living for different groups can be constructed by compiling weights for each group. These weights require cross-section data like those collected in the Survey of Consumer Expenditures.

By contrast, the econometric approach is based on sample moments of cross-section data on individual expenditures on specific commodities, total expenditures and the demographic characteristics of individual households. These sample moments are combined with time series data on prices and aggregate expenditure patterns to estimate the parameters of our econometric model. Estimates like those presented in Section 3 summarize the cross-section and time series data in a concise and readily intelligible way.

We have found it convenient to employ price data for commodity groups from the U.S. national accounts. These data are based largely on price information compiled for the Consumer Price Index published by the Bureau of Labor Statistics. Our approach could be implemented for price data taken directly from the Consumer Price Index. This would have the advantage of providing greater comparability between the results of the econometric approach and the index number approach employed currently.

Our econometric model employs time series data on the level of aggregate expenditure, its distribution over commodity groups, and its distribution over the population and among demographic groups. We have constructed these data from the Current Population Survey, which is conducted monthly in the United States. The necessary information is contained in the summary statistics of the joint distribution of expenditures and attributes -- ΣM_k $\ln M_k / M$ and $\left\{ \Sigma M_k A_k / M \right\}$ -- presented in Section 2.

Although the econometric approach to cost-of-living measurement requires more data than the conventional index number approach, the econometric approach has overwhelming practical advantages. First and foremost, it allows for substitution among commodities in response to price changes. To achieve similar results by means of conventional index numbers, it would be necessary to have surveys of consumer expenditures at the same frequency as the intervals used for reporting the cost-of-living index, which would be far too burdensome.

A second advantage of the econometric approach to cost-of-living measurement is in flexibility and ease of application. We have implemented a social cost-of-living index for the United States in Section 6. In addition, we have compared cost-of-living increases for typical consuming units in Section 4. Finally, we have made comparisons among cost-of-living increases for groups of consuming units in Section 7. Our approach could also be used to construct special cost-of-living indexes for groups such as low income households receiving government transfer payments or aged households living on pensions.

While the econometric approach has important advantages over the conventional index number approach, it is important to emphasize that the two approaches share a number of limitations. In principle prices for commodity groups must be compiled for goods and services of constant quality. In actuality statisticians are faced with rapidly changing commodity specifications and with the introduction of new commodities. As an illustration, the addition of pollution control equipment to automobiles poses exactly the same problems for the two approaches.

A second set of problems common to the econometric and the index number approaches is posed by consumption not associated with direct purchases of goods and services. As an illustration, the imputed value of home produced goods such as food produced on farms must be included in total expenditure. Similarly, an imputation is required for flows of services from owner-occupied housing and owner-utilized transportation equipment and other consumer durables. A more complex range of problems is posed by goods and services consumed collectively, such as police protection and environmental quality.

Finally, the implementation of the econometric approach described in this paper has important limitations of its own. We have employed prices for all commodity groups compiled at the national level. A more detailed implementation incorporating regional and other differences in prices actually paid would be useful in many applications. In addition, it would be very desirable to provide a more detailed commodity breakdown. These limitations can be overcome by expenditure of greater resources of human effort and computer time. Fortunately, existing data bases will be adequate for the construction of far more detailed econometric models than the model we have presented.

Up to this point we have compared the econometric approach with the conventional index number approach, such as that employed by the Bureau of Labor Statistics in the United States or by Statistics Canada. It is also interesting to compare the econometric approach with a more sophisticated index number approach proposed by Diewert [1976, 1981]. Diewert's approach is based on exact index numbers.[24] Exact index numbers do not require an econometric model, but reproduce exactly the individual cost-of-living index derived from an individual expenditure function.

An important example of the sophisticated index number approach is the exact translog cost-of-living index considered by Tornqvist [1936]:

$$\ln P_k(p^1, p^0, V_k^*) = \bar{w}_k' \, \Delta \ln p, \qquad (k = 1, 2 \dots K), \qquad (8.1)$$

where V_k^* is the base level of individual welfare:

$$V_k^* = (V_k^1 V_k^0)^{\frac{1}{2}}, \qquad (k = 1, 2 \dots K),$$

and:

$$\bar{w}_k = \frac{1}{2}(w_k^1 + w_k^0), \qquad (k = 1, 2 \dots K),$$

$$\Delta \ln p = \ln p^1 - \ln p^0.$$

Diewert [1976, 1981] has shown that this cost-of-living index is exact for a translog individual expenditure function similar but not identical to our expenditure function (2.23).

The data required for implementation of the exact index number approach are far more extensive than those required for the econometric approach. For example, the exact translog cost-of-living index (8.1) would require data on individual expenditure shares for each period in which a cost-of-living comparison is needed. An annual time series of comparisons would require that a panel of consumers would have to be surveyed annually.

It is important to note that the exact index number approach is not limited to individual cost-of-living measurement. For social or group cost-of-living measurement a translog cost-of-living index could be defined on the basis of the corresponding translog social or group expenditure functions. The individual budget shares in the index (8.1) could be replaced by aggregate or group expenditure shares.

In order to apply the exact translog cost-of-living index (8.1) to society as a whole, it would be necessary to obtain aggregate expenditure shares from an econometric model. These shares would correspond to those of a representative consumer, constructed by equalizing aggregate expenditure per household equivalent member over all consuming units. Data on aggregate expenditure shares employed in fitting an econometric model would not be appropriate for this purpose. These expenditure shares correspond to the actual distribution of total expenditure over the population.

Similarly, the exact translog cost-of-living index (8.1) could be applied to groups, but only by calculating expenditure shares for the group for an econometric model. As before, these shares would correspond to those of a representative consumer. The demand system for the representative consumer would be obtained by equalizing group expenditure per household equivalent member over the households included in the group.

Despite the limitations of the exact index number approach, we find it useful to present an exact translog social cost-of-living index calculated from the Tornqvist formula (8.1) in Table 5. This index can be compared with the translog social cost-of-living index presented in Table 3. The two indexes are similar, but not identical. An important difference between the two is that the exact translog index is a chain index with base welfare levels changing from period to period, while the translog index presented in Table 3 employs the level of welfare in 1972 as a base.

Our final conclusion is that the econometric approach has very substantial advantages over both conventional and sophisticated index number approaches. However, the econometric approach is not a panacea for the solution of all the practical problems of cost-of-living measurement. As better solutions to these problems become available, the results can be incorporated into an econometric model like that we have presented.

TABLE 5. Exact Translog Cost-of-Living Index

Year	Exact Cost-of-Living Index (1972 = 1.0000)	Inflation Rate ((annual percentage rates)
1958	.6934	----
1959	.6952	0.26
1960	.7156	2.89
1961	.7216	0.83
1962	.7293	1.06
1963	.7334	0.56
1964	.7490	2.10
1965	.7730	3.15
1966	.7939	2.67
1967	.8018	0.99
1968	.8295	3.40
1969	.8823	6.17
1970	.8981	1.77
1971	.9353	4.06
1972	1.0000	6.69
1973	1.0755	7.28
1974	1.1632	7.84
1975	1.2310	5.66
1976	1.3142	6.54
1977	1.4330	8.65
1978	1.5254	6.25

APPENDIX TABLE 1. Individual Cost-of-Living Indexes (1972 = 1.0000)

Year	Urban		Rural	
	White	Nonwhite	White	Nonwhite
1958	0.6969	0.6968	0.7002	0.7001
1959	0.6967	0.6964	0.7019	0.7016
1960	0.7176	0.7173	0.7209	0.7207
1961	0.7236	0.7233	0.7272	0.7269
1962	0.7311	0.7308	0.7346	0.7342
1963	0.7345	0.7341	0.7392	0.7389
1964	0.7508	0.7507	0.7535	0.7534
1965	0.7758	0.7756	0.7760	0.7758
1966	0.7992	0.7989	0.7998	0.7994
1967	0.8051	0.8048	0.8086	0.8084
1968	0.8310	0.8311	0.8353	0.8354
1969	0.8840	0.8845	0.8849	0.8854
1970	0.8973	0.8973	0.9042	0.9042
1971	0.9319	0.9317	0.9390	0.9388
1972	1.0000	1.0000	1.0000	1.0000
1973	1.0856	1.0848	1.0834	1.0826
1974	1.1864	1.1818	1.2014	1.1968
1975	1.2430	1.2375	1.2681	1.2625
1976	1.3251	1.3197	1.3479	1.3425
1977	1.4407	1.4341	1.4590	1.4524
1978	1.5459	1.5385	1.5633	1.5558

Region = Northeast.
Size = 5.
Age = 35-44.
Expenditure = $4,467.

D.W. Jorgenson and D.T. Slesnick

APPENDIX TABLE 1. Individual Cost-of-Living Indexes (1972 = 1.0000) - Continued

	Urban		Rural	
Year	White	Nonwhite	White	Nonwhite
1958	0.6946	0.6945	0.6978	0.6978
1959	0.6952	0.6949	0.7004	0.7001
1960	0.7162	0.7159	0.7195	0.7193
1961	0.7222	0.7219	0.7258	0.7255
1962	0.7299	0.7296	0.7334	0.7330
1963	0.7335	0.7332	0.7382	0.7379
1964	0.7500	0.7498	0.7527	0.7525
1965	0.7742	0.7740	0.7745	0.7743
1966	0.7967	0.7964	0.7973	0.7970
1967	0.8037	0.8034	0.8072	0.8069
1968	0.8304	0.8305	0.8348	0.8349
1969	0.8836	0.8841	0.8845	0.8850
1970	0.8969	0.8970	0.9039	0.9039
1971	0.9328	0.9326	0.9398	0.9397
1972	1.0000	1.0000	1.0000	1.0000
1973	1.0802	1.0794	1.0780	1.0771
1974	1.1720	1.1675	1.1868	1.1823
1975	1.2282	1.2228	1.2530	1.2475
1976	1.3145	1.3092	1.3372	1.3318
1977	1.4309	1.4244	1.4491	1.4425
1978	1.5301	1.5228	1.5473	1.5399

Region = Northeast.
Size = 5.
Age = 35-44.
Expenditure = $8,934.

APPENDIX TABLE 1. Individual Cost-of-Living Indexes (1972 = 1.0000) - Concluded

Year	Urban		Rural	
	White	Nonwhite	White	Nonwhite
1958	0.6923	0.6922	0.6955	0.6954
1959	0.6937	0.6934	0.6988	0.6986
1960	0.7148	0.7145	0.7181	0.7179
1961	0.7209	0.7206	0.7244	0.7241
1962	0.7287	0.7284	0.7322	0.7318
1963	0.7326	0.7322	0.7373	0.7369
1964	0.7491	0.7490	0.7518	0.7517
1965	0.7727	0.7724	0.7729	0.7727
1966	0.7943	0.7939	0.7948	0.7945
1967	0.8023	0.8020	0.8058	0.8055
1968	0.8298	0.8299	0.8342	0.8343
1969	0.8833	0.8837	0.8842	0.8846
1970	0.8966	0.8966	0.9035	0.9035
1971	0.9336	0.9335	0.9407	0.9405
1972	1.0000	1.0000	1.0000	1.0000
1973	1.0748	1.0740	1.0725	1.0717
1974	1.1577	1.1533	1.1724	1.1679
1975	1.2137	1.2083	1.2382	1.2327
1976	1.3041	1.2988	1.3266	1.3212
1977	1.4212	1.4147	1.4392	1.4327
1978	1.5145	1.5073	1.5315	1.5242

Region = Northeast.
Size = 5.
Age = 35-44.
Expenditure = $17,864.

APPENDIX TABLE 2. Group Cost-of-Living Indexes (1972 = 1.0000)

	Family Size						
Year	1	2	3	4	5	6	7+
1958	0.6917	0.6927	0.6930	0.6930	0.6932	0.6935	0.6932
1959	0.6906	0.6946	0.6952	0.6953	0.6956	0.6958	0.6956
1960	0.7147	0.7157	0.7160	0.7158	0.7156	0.7157	0.7148
1961	0.7204	0.7219	0.7221	0.7219	0.7218	0.7220	0.7212
1962	0.7283	0.7298	0.7301	0.7299	0.7297	0.7299	0.7291
1963	0.7299	0.7336	0.7341	0.7341	0.7343	0.7344	0.7340
1964	0.7487	0.7498	0.7500	0.7499	0.7497	0.7498	0.7492
1965	0.7764	0.7733	0.7730	0.7726	0.7720	0.7722	0.7710
1966	0.7971	0.7947	0.7944	0.7942	0.7941	0.7945	0.7940
1967	0.8004	0.8029	0.8033	0.8033	0.8036	0.8038	0.8037
1968	0.8252	0.8300	0.8308	0.8311	0.8316	0.8317	0.8320
1969	0.8830	0.8831	0.8832	0.8832	0.8831	0.8831	0.8828
1970	0.8864	0.8965	0.8979	0.8987	0.9003	0.9004	0.9021
1971	0.9232	0.9340	0.9354	0.9361	0.9374	0.9373	0.9386
1972	1.0000	1.0000	1.0000	1.0000	1.0000	1.0000	1.0000
1973	1.0798	1.0741	1.0734	1.0736	1.0744	1.0752	1.0764
1974	1.1493	1.1600	1.1610	1.1624	1.1672	1.1692	1.1742
1975	1.1921	1.2168	1.2195	1.2219	1.2288	1.2309	1.2382
1976	1.2877	1.3090	1.3111	1.3123	1.3169	1.3184	1.3224
1977	1.4161	1.4283	1.4290	1.4287	1.4310	1.4325	1.4333
1978	1.5125	1.5208	1.5210	1.5211	1.5247	1.5270	1.5295

APPENDIX TABLE 2. Group Cost-of-Living Indexes (1972 = 1.0000) - Continued

	Age of Head					
Year	16-24	25-34	35-44	45-54	55-64	65 +
1958	0.6930	0.6938	0.6926	0.6926	0.6928	0.6927
1959	0.6911	0.6940	0.6947	0.6952	0.6956	0.6943
1960	0.7166	0.7165	0.7153	0.7152	0.7156	0.7154
1961	0.7221	0.7224	0.7214	0.7215	0.7219	0.7216
1962	0.7299	0.7301	0.7293	0.7295	0.7299	0.7295
1963	0.7305	0.7328	0.7335	0.7341	0.7345	0.7332
1964	0.7500	0.7503	0.7495	0.7494	0.7497	0.7495
1965	0.7792	0.7758	0.7723	0.7717	0.7719	0.7734
1966	0.7990	0.7967	0.7938	0.7938	0.7939	0.7952
1967	0.8008	0.8026	0.8029	0.8034	0.8036	0.8028
1968	0.8246	0.8288	0.8308	0.8312	0.8314	0.8297
1969	0.8839	0.8842	0.8834	0.8826	0.8826	0.8828
1970	0.8828	0.8923	0.8984	0.9000	0.9003	0.8962
1971	0.9198	0.9290	0.9357	0.9374	0.9377	0.9333
1972	1.0000	1.0000	1.0000	1.0000	1.0000	1.0000
1973	1.0802	1.0769	1.0730	1.0741	1.0738	1.0763
1974	1.1448	1.1550	1.1585	1.1671	1.1680	1.1644
1975	1.1827	1.2048	1.2170	1.2290	1.2304	1.2210
1976	1.2823	1.2982	1.3076	1.3177	1.3194	1.3114
1977	1.4159	1.4207	1.4236	1.4328	1.4349	1.4306
1978	1.5119	1.5154	1.5154	1.5261	1.5278	1.5254

APPENDIX TABLE 2. Group Cost-of-Living Indexes (1972 = 1.0000) - Concluded

Year	NE	NC	S	W	White	Non-White	Urban	Rural
1958	0.6927	0.6932	0.6936	0.6910	0.6927	0.6942	0.6927	0.6953
1959	0.6937	0.6953	0.6964	0.6919	0.6945	0.6956	0.6943	0.6997
1960	0.7152	0.7161	0.7162	0.7140	0.7155	0.7162	0.7154	0.7184
1961	0.7213	0.7222	0.7225	0.7200	0.7216	0.7223	0.7215	0.7248
1962	0.7292	0.7302	0.7304	0.7280	0.7296	0.7300	0.7294	0.7327
1963	0.7325	0.7342	0.7351	0.7312	0.7335	0.7340	0.7332	0.7383
1964	0.7493	0.7501	0.7503	0.7484	0.7496	0.7501	0.7495	0.7521
1965	0.7738	0.7732	0.7722	0.7732	0.7730	0.7735	0.7731	0.7722
1966	0.7954	0.7947	0.7942	0.7944	0.7946	0.7956	0.7947	0.7940
1967	0.8024	0.8034	0.8041	0.8011	0.8029	0.8037	0.8028	0.8063
1968	0.8291	0.8308	0.8322	0.8278	0.8302	0.8311	0.8300	0.8351
1969	0.8831	0.8833	0.8833	0.8826	0.8830	0.8838	0.8831	0.8838
1970	0.8949	0.8978	0.9010	0.8930	0.8971	0.8983	0.8967	0.9055
1971	0.9317	0.9352	0.9383	0.9304	0.9344	0.9346	0.9339	0.9434
1972	1.0000	1.0000	1.0000	1.0000	1.0000	1.0000	1.0000	1.0000
1973	1.0770	1.0740	1.0733	1.0753	1.0745	1.0769	1.0750	1.0701
1974	1.1612	1.1623	1.1674	1.1525	1.1614	1.1672	1.1612	1.1737
1975	1.2154	1.2207	1.2302	1.2044	1.2188	1.2252	1.2179	1.2432
1976	1.3060	1.3119	1.3192	1.2974	1.3099	1.3130	1.3087	1.3324
1977	1.4256	1.4299	1.4336	1.4189	1.4281	1.4284	1.4270	1.4446
1978	1.5209	1.5225	1.5262	1.5119	1.5211	1.5244	1.5206	1.5346

Footnotes

[1] The literature on the individual cost-of-living index is summarized by Deaton and Muellbauer [1980], pp.170-178, and by Diewert [1981].

[2] The translog indirect utility function was introduced by Christensen, Jorgenson and Lau [1975] and extended to encompass determinants of expenditure allocation other than prices and total expenditure by Jorgenson and Lau [1975]. Alternative approaches to the representation of the effects of prices and total expenditure on expenditure allocation are summarized by Barten [1977], Deaton and Muellbauer [1980], pp.60-85, and Lau [1977a].

[3] Alternative approaches to the representation of household characteristics on expenditure allocation are presented by Barten [1964], Gorman [1976], and Prais and Houthakker [1971]. Empirical evidence on the impact of demographic characteristics on expenditure allocation is given by Lau, Lin and Yotopoulos [1978], Muellbauer [1977], Parks and Barten [1973], Pollak and Wales [1980, 1981], and Ray [1982].

[4] The specification of a system of individual demand functions by means of Roy's Identity was first employed in econometric modeling of consumer behavior by Houthakker [1960]. A detailed review of econometric models based on Roy's Identity is given by Lau [1977a].

[5] For further discussion, see Lau [1977b, 1982] and Jorgenson, Lau and Stoker [1980, 1981, 1982].

[6] Conditions for integrability are discussed by Jorgenson and Lau [1979] and by Jorgenson, Lau and Stoker [1982].

[7] For further discussion see Jorgenson, Lau and Stoker [1982], esp. pp.175-186.

[8] Household equivalence scales are discussed by Barten [1964], Lazear and Michael [1980], Muellbauer [1974, 1977, 1980], and Prais and Houthakker [1971], among others. Alternative approaches are summarized by Deaton and Muellbauer [1980].

[9] The use of household equivalence scales in evaluating transfers among individuals has been advocated by Deaton and Muellbauer [1980], esp. pp.205-212, and by Muellbauer [1974]. Pollak and Wales [1979] have presented arguments against the use of household equivalence scales for this purpose.

[10] The 1972-1973 Survey of Consumer Expenditures is discussed by Carlson [1974].

[11] We employ data on the flow of services from durable goods rather than purchases of durable goods. Personal consumption expenditures in the U.S. National Income and Product Accounts are based on purchases of durable goods.

[12] This series is published annually by the U.S. Bureau of the Census.

[13] A detailed discussion of the stochastic specification of our model and of econometric methods for pooling time series and cross-section data is presented by Jorgenson, Lau and Stoker [1982], Section 6. This stochastic specification implies that time series data must be adjusted for

heteroscedasticity by multiplying each observation by the statistic:

$$\rho = \frac{(\Sigma \ M_k)^2}{\Sigma \ M_k^2}.$$

[14] An alterative approach to implementation of the individual cost-of- living index (4.2) is to bound this index on the basis of observable data. Bounds have been developed by Allen [1949], Frisch [1936] and Konüs [1939] and, more recently, by Afriat [1977], Pollak [1971, 1981], and Samuelson and Swamy [1974], among many others. This approach is reviewed by Deaton and Muellbauer [1980], pp.170-178, and by Diewert [1981].

[15] Arrow [1977, p.225] has defended noncomparability in the following terms: ... the autonomy of individuals, an element of mutual incommensurability among people seems denied by the possibility of interpersonal comparisons.

[16] It is important to note that the social welfare function is (5.2) represents a social ordering over all possible individual orderings and exemplifies the multiple profile approach to social choice Arrow [1963] rather than the single profile approach employed by Bergson [1938] and Samuelson [1947]. The literature on the existence of single profile social welfare functions is discussed by Roberts [1980d], Samuelson [1982] and Sen [1979b].

[17] See Sen [1977, 1979b] for further discussion.

[18] See Roberts [1980b], esp. pp.434-436.

[19] The implications of distributional homotheticity are discussed by Kolm [1976b] and Blackorby and Donaldson [1978].

[20] Mean value functions were introduced into economics by Bergson [1936] and have been employed, for example, by Arrow, Chenery, Minhas and Solow [1961] and Atkinson [1970]. Properties of mean value functions are discussed by Hardy, Littlewood and Polya [1959].

[21] This assumption implies that individual welfare increases with total expenditure at a rate that is inversely proportional to total expenditure. This is also implied by the utilitarian social welfare function employed by Arrow and Kalt [1979].

[22] The social expenditure function was introduced by Pollak [1981]. Alternative money measures of social welfare are discussed by Arrow and Kalt [1979], Bergson [1980], Deaton and Muellbauer [1980], pp.214-239, Roberts [1980c], and Sen [1976]. A survey of the literature is presented by Sen [1979a].

[23] The conventional index number approach is summarized in Statistics Canada [1982].

[24] Exact index numbers are discussed by Lau [1979], Pollak [1971] and Samuelson and Swamy [1974].

References

Afriat, S.N. [1977], *The Price Index*, London, Cambridge University Press.

Allen, R.G.D. [1949], "The Economic Theory of Index Numbers", *Economica*, Vol. 16, No. 63, August, pp.197-203.

Arrow, K.J. [1963], *Social Choice and Individual Values*, New Haven, Yale University Press, 2nd ed.

_____ [1977], "Extended Sympathy and the Possibility of Social Choice", *American Economic Review*, Vol. 67, No. 1, February, pp.219-225.

_____ H.B. Chenery, B.S. Minhas and R.M. Solow [1961], "Capital-Labor Substitution and Economic Efficiency", *Review of Economics and Statistics*, Vol. 43, No. 3, August, pp.225-250.

_____ and J.P. Kalt [1979], *Petroleum Price Regulation: Should We Decontrol*, Washington, American Enterprise Institute.

d'Aspremont, C. and L. Gevers [1977], "Equity and the Informational Basis of Collective Choice", *Review of Economic Studies*, Vol. 44, No. 137, June, pp.199-209.

Atkinson, A.B. [1970], "On Measurement of Inequality", *Journal of Economic Theory*, Vol. 2, No. 3, September, pp.244-263.

Barten, A.P. [1964], "Family Composition, Prices, and Expenditure Patterns", in P. Hart, G. Mills and J.K. Whitaker (eds.), *Econometric Analysis for National Economic Planning: 16th Symposium of the Colston Society*, London, Butterworth, pp.277-292.

_____ [1977], "The Systems of Consumer Demand Functions Approach: A Review", in M.D. Intriligator (ed.), *Frontiers of Quantitative Economics*, Vol. IIIA, Amsterdam, North-Holland, pp.23-58.

Bergson, A. [1936], "Real Income, Expenditure Proportionality, and Frisch's 'New Methods of Measuring Marginal Utility'', *Review of Economic Studies*, Vol. 4, No. 1, October 1936, pp.33-52.

_____ [1938], "A Reformulation of Certain Aspects of Welfare Economics", *Quarterly Journal of Economics*, Vol. 52, No. 2, February, pp.310-334.

_____ [1980], "Consumer's Surplus and Income Redistribution", *Journal of Public Economics*, Vol. 14, No. 1, August, pp.31-47.

Blackorby, C. and D. Donaldson [1978], "Measures of Relative Equality and their Meaning in Terms of Social Welfare", *Journal of Economic Theory*, Vol. 18, No. 1, June, pp.651-675.

_____ [1982], "Ratio-Scale and Translation-Scale Full Interpersonal Comparability without Domain Restrictions: Admissible Social-Evaluation Functions", *International Economic Review*, Vol. 23, No. 2, June, pp.249-268.

Bureau of the Census (various annual issues), *Current Population Reports, Consumer Income, Series P-60*, Washington, D.C., U.S. Department of Commerce.

Carlson, M.D. [1974], "The 1972-73 Consumer Expenditure Survey", *Monthly Labor Review*, Vol. 97, No. 12, December, pp.16-23.

Christensen, L.R., D.W. Jorgenson and L.J. Lau [1975], "Transcendental Logarithmic Utility Functions", *American Economic Review*, Vol. 65, No. 3, June, pp.367-383.

Dalton, H. [1920], "The Measurement of Inequality of Income", *Economic Journal*, Vol. 30, No. 119, September, pp.361-384.

Deaton, A. and J. Muellbauer [1980], *Economics and Consumer Behavior*, Cambridge, Cambridge University Press.

Deschamps, R. and L. Gevers [1978], "Leximin and Utilitarian Rules: A Joint Characterisation", *Journal of Economic Theory*, Vol. 17, No. 2, April, pp.143-163.

Diewert, W.E. [1976], "Exact and Superlative Index Numbers", *Journal of Econometrics*, Vol. 4, No. 2, May, pp.115-145.

——————— [1981], "The Economic Theory of Index Numbers: A Survey", in A. Deaton (ed.), *Essays on Theory and Measurement of Consumer Behavior in Honor of Sir Richard Stone*, pp.163-208.

Frisch, R. [1936], "Annual Survey of Economic Theory: The Problem of Index Numbers", *Econometrica*, Vol. 4, No. 1, January, pp.1-39.

Gorman, W.M. [1976], "Tricks with Utility Functions", in M.J. Artis and A.R. Nobay (eds.), *Essays in Economic Analysis*: *Proceedings of the 1975 AUTE Conference*, Cambridge, Cambridge University Press, pp.211-243.

Hammond, P.J. [1976], "Equity, Arrow's Conditions and Rawl's Difference Principle", *Econometrica*, Vol. 44, No. 4, July, pp.793-804.

——————— [1977], "Dual Interpersonal Comparisons of Utility and the Welfare Economics of Income Distribution", *Journal of Public Economics*, Vol. 7, No. 1, February, pp.51-71.

Hardy, G.H., J.E. Littlewood and G. Polya [1959], *Inequalities*, Cambridge, Cambridge University Press, 2nd ed.

Harsanyi, J.C. [1976], *Essays on Ethics, Social Behavior and Scientific Explanation*, Dordrecht, D. Reidel.

Houthakker, H.S. [1960], "Additive Preferences", *Econometrica*, Vol. 28, No. 2, April, pp.244-257.

Jorgenson, D.W. and L.J. Lau [1975], "The Structure of Consumer Preferences", *Annals of Economic and Social Measurement*, Vol. 4, No. 1, January, pp.49-101.

_____ and _____ [1979], "The Integrability of Consumer Demand Functions", *European Economic Review*, Vol. 12, No. 2, April, pp.115-147.

_____ L.J. Lau and T.M. Stoker [1980], "Welfare Comparison Under Exact Aggregation", *American Economic Review*, Vol. 70, No. 2, May, pp.268-272.

_____ , _____ , and _____ [1981], "Aggregate Consumer Behavior and Individual Welfare", in D. Currie, R. Nobay and D. Peel (eds.), *Macroeconomic Analysis*, London, Croom-Helm, pp.35-61.

_____ [1982], "The Transcendental Logarithmic Model of Aggregate Consumer Behavior", in R.L. Basmann and G.F. Rhodes Jr. (eds.), *Advances in Econometrics*, Vol. 1, Greenwich, JAI Press, pp.97-238.

Kolm, S.C. [1969], "The Optimal Production of Social Justice", in J. Margolis and H. Guitton (eds.), *Public Economics*, London: Macmillan, pp.145-200.

_____ [1976a and 1976b], "Unequal Inequalities I and II", *Journal of Economic Theory*, Vol. 12, No. 3, June, pp.416-42, and Vol. 13, No. 1, August, pp.82-111.

Konüs, A.A. [1939], "The Problem of the True Index of the Cost-of-Living", *Econometrica*, Vol. 7, No. 1, January, pp.10-29; originally published in 1924.

Lau, L.J. [1977a], "Complete Systems of Consumer Demand Functions through Duality", in M.D. Intriligator (ed.), *Frontiers of Quantitative Economics*, Vol. IIIA, Amsterdam, North-Holland, pp.59-86.

_____ [1977b], "Existence Conditions for Aggregate Demand Functions", Technical Report No. 248, Institute for Mathematical Studies in the Social Sciences, Stanford University, Stanford (revised 1980 and 1982).

_____ [1979], "On Exact Index Numbers", *Review of Economics and Statistics*, Vol. 61, No. 1, February, pp.73-82.

_____ [1982], "A Note on the Fundamental Theorem of Exact Aggregation", *Economics Letters*, Vol. 9, No. 2, pp.119-126.

Lau, L.J., W.L. Lin and P.A. Yotopoulos [1978], "The Linear Logarithmic Expenditure System: An Application to Consumption-Leisure Choice", *Econometrica*, Vol. 46, No. 4, July, pp.843-868.

Lazear, E.P. and R.T. Michael [1980], "Family Size and the Distribution of Real Per Capita Income", *American Economic Review*, Vol. 70, No. 1, March, pp.91-107.

Martos, B. [1969], "Subdefinite Matrices and Quadratic Forms", *SIAM Journal of Applied Mathematics*, Vol. 17, pp.1215-1223.

Maskin, E. [1978], "A Theorem on Utilitarianism", *Review of Economic Studies*, Vol. 42, No. 139, February, pp.93-96.

Mirrlees, J.A. [1971], "An Exploration in the Theory of Optimal Income Taxation", *Review of Economic Studies*, Vol. 38, No. 114, April, pp.175-208.

Muellbauer, J. [1974], "Household Composition, Engel Curves and Welfare Comparisons between Households: A Duality Approach", *European Economic Review*, Vol. 5, No. 2, August, pp.103-122.

_____ [1977], "Testing the Barten Model of Household Composition Effects and the Cost of Children", *Economic Journal*, Vol. 87, No. 347, September, pp. 460-487.

_____ [1980], "The Estimation of the Prais-Houthakker Model of Equivalence Scales", *Econometrica*, Vol. 48, No. 1, January, pp. 153-176.

Ng, Y.K. [1975], "Bentham or Bergson? Finite Sensibility, Utility Functions and Social Welfare Functions", *Review of Economic Studies*, Vol 42, No. 4, October, pp. 545-569.

Parks, R.W. and A.P. Barten [1973], "A Cross Country Comparison of the Effects of Prices, Income, and Population Composition on Consumption Patterns", *Economic Journal*, Vol. 83, No. 331, September, pp. 834-852.

Pollak, R.A. [1971], "The Theory of the Cost of Living Index", Research Discussion Paper No. 11, Office of Prices and Living Conditions, U.S. Bureau of Labor Statistics, Washington, D.C.

————— [1981], "The Social Cost of Living Index", *Journal of Public Economics*, Vol. 15, No. 3, June, pp. 311-336.

Pollak, R.A. and T.J. Wales [1979], "Welfare Comparisons and Equivalent Scales", *American Economic Review*, Vol. 69, No. 2, May, pp. 216-21.

————— , ————— , and ————— [1980] "Comparisons of the Quadratic Expenditure System and Translog Demand Systems with Alternative Specifications of Demographic Effects", *Econometrica*, Vol. 48, No. 3, April, pp. 595-612.

————— , ————— , and ————— [1981] "Demographic Variables in Demand Analysis", *Econometrica*, Vol. 49, No. 6, November, pp. 1533-1552.

Prais, S.J. and H.S. Houthakker [1971], *The Analysis of Family Budgets*, Cambridge, Cambridge University Press, 2nd ed.

Ray, R. [1982], "The Testing and Estimation of Complete Demand Systems on Household Budget Surveys: An Application of AIDS", *European Economic Review*, Vol. 17, No. 3, March 1982, pp. 349-370.

Roberts, K.W.S. [1980a], "Possibility Theorems with Interpersonally Comparable Welfare Levels", *Review of Economic Studies*, Vol. 47, No. 147, January, pp. 409-20.

————— [1980b], "Interpersonal Comparability and Social Choice Theory", *Review of Economic Studies*, Vol. 47, No. 147, January, pp. 421-439.

_____ [1980c], "Price-Independent Welfare Prescriptions", *Journal of Public Economics*, Vol. 13, No. 3, June, pp. 277-298.

_____ [1980d], "Social Choice Theory: The Single-profile and Multi-profile Approaches", *Review of Economic Studies*, Vol. 47, No. 147, January, pp. 441-450.

Roy, R. [1943], *De l'Utilite: Contributions a la Theorie des Choix*, Paris, Herman.

Samuelson, P.A. [1947], *Foundations of Economic Analysis*, Cambridge, Harvard University Press.

_____ [1956], "Social Indifference Curves", *Quarterly Journal of Economics*, Vol. 70, No. 1, February, pp. 1-22.

_____ [1982], "Bergsonian Welfare Economics", in S. Rosefielde (ed.), *Economic Welfare and the Economics of Soviet Socialism: Essays in Honor of Abram Bergson*, Cambridge, Cambridge University Press, pp. 223-266.

_____ and S. Swamy [1974], "Invariant Economic Index Numbers and Canonical Duality: Survey and Synthesis", *American Economic Review*, Vol. 64, No. 4, September, pp. 566-93.

Sen, A.K. [1970], *Collective Choice and Social Welfare*, Edinburgh, Oliver and Boyd.

_____ [1973], *On Economic Inequality*, Oxford, Clarendon Press.

_____ [1976], "Real National Income", *Review of Economic Studies*, Vol. 43, No. 133, February, pp.19-40.

_____ [1977], "On Weights and Measures: Informational Constraints in Social Welfare Analysis", *Econometrica,* Vol. 45, No. 7, October, pp.1539-1572.

_____ [1979a], "The Welfare Basis of Real Income Comparisons: A Survey", *Journal of Economic Literature*, Vol. 17, No. 1, March, pp.1-45.

_____ [1979b], "Personal Utilities and Public Judgements: Or What's Wrong with Welfare Economics", *Economic Journal*, Vol. 89, No. 763, September, pp.537-558.

Statistics Canada [1982], *The Consumer Price Index Reference Paper*, Ottawa, Statistics Canada, May.

Tornqvist, L. [1936], "The Bank of Finland's Consumption Price Index", *Bank of Finland Monthly Bulletin*, Vol. 10, pp.1-8.

PRICE LEVEL MEASUREMENT – W.E. Diewert (Editor)
Canadian Government Publishing Centre /
Elsevier Science Publishers B.V. (North-Holland)
© Minister of Supply and Services Canada, 1990

COMMENTS

R. Robert Russell
Department of Economics
New York University

My professional life-long fear has been that one day Erwin Diewert would organize a conference and make me discuss a paper by Dale Jorgenson. It's a little like being asked to give a 10-minute critique of the *Encyclopedia of Economics*.

In this paper, Jorgenson and Slesnick review consumer theory – from integrability conditions to duality, the theory and application of consumer equivalence scales, the theory of aggregation, the econometrics of pooled time-series and cross-section data, the theory of cost-of-living indexes, and the theory of social choice and welfare economics – from Arrow's impossibility theorem right up to recent results on interpersonal comparability and cardinal utility. Another tour de force.

But the paper is much more than review; it is an invaluable contribution. Especially noteworthy is the approach to the econometric estimation of market demand functions obtained by aggregating demand functions of individuals with different tastes, incorporating the restrictions implied by the theory of individual consumer behaviour. As this approach has been reported before,[1] I concentrate my remarks on the other important contribution: the theory and application of social cost-of-living indexes derived from social welfare functions and individual demand functions. This is the first empirical application (of which I am aware) of the theory of social cost-of-living indexes.

To put the Jorgenson-Slesnick (J-S) approach in perspective, let me first set up the social cost-of-living-index problem in the manner of Pollak [1981]. The expenditure function of the rth individual, E^r, is defined by

$$E^r(u_r, P) = \min_{X^r} \left\{ P \cdot X^r \mid U^r(X^r) \geq u_r \right\}, \tag{1}$$

where u_r, X^r, and U^r are the utility level, consumption bundle, and utility function, respectively, of the rth consumer, and P is the price vector. The aggregate expenditure function

E, conditional on the utility profile, $u = u_1,...,u_R$, is defined by

$$E(u_1,...,u_R,P) = \sum_r E^r(u_r,P). \qquad (2)$$

For a fixed aggregate expenditure, M, and a fixed P, the equality,

$$E(u_1,...,u_R,P) = M, \qquad (3)$$

defines a utility possibility frontier, labelled UPF(P,M). The utility possibility frontier represents the maximal utility for each individual given the utilities of all other individuals (and given M and P). It is illustrated for the case of two individuals in Figure 1.

The utility possibility frontier presumes individual optimization (i.e., efficiency) only; no elements of cardinal utility or interpersonal utility comparisons are involved. It shifts upward with increases in M and downward with increases in P; it is the social analogue of the individual budget constraint.

Social optimality – the choice of a point on the utility possibility frontier – requires the specification of a social welfare function, W. For a particular value of social welfare, ω, the equality,

$$W(u) = W(u_1,...,u_R) = \omega, \qquad (4)$$

defines a social indifference curve (SIC(ω)). Social optimality (for given M and P) is represented by the tangency of a SIC(ω) and the UPF(P,M) (at u^* in Figure 1).

The social expenditure function, ξ, is defined by

$$\xi(\omega,P) = \min_{u_1,...,u_R} \left\{E(u_1,...,u_R,P) \mid W(u) \geq \omega\right\}. \qquad (5)$$

The Pollak social cost-of-living index, applied by Jorgenson and Slesnick, is

Figure 1

Figure 2

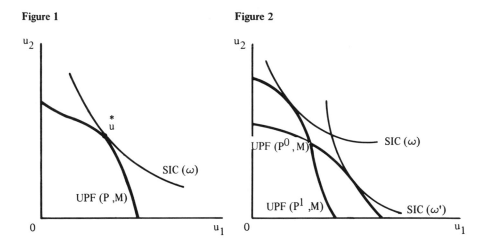

Figure 3

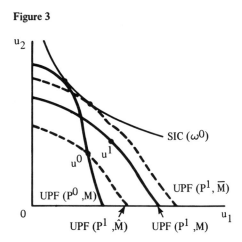

$$SCOL(\omega, P^0, P^1) = \frac{\xi(\omega, P^1)}{\xi(\omega, P^0)} \cdot \tag{6}$$

There are two fundamental problems with this approach. To illustrate the first, consider a change in prices from P^0 to P^1, holding M constant, with corresponding utility possibility frontiers reflected in Figure 2. Consider alternative social welfare functions, W and W', with social indifference curves labelled SIC(ω) and SIC (ω'), respectively. Given the social welfare function, W, the cost of living is lower in situation 0 than in situation 1, since M would have to be increased to raise UPF(P^0,M) to a point of tangency with SIC(ω). Given the social welfare function W', the cost of living is **higher** in situation 0 than in situation 1, since M must be increased to obtain a tangency of UPF(P^1,M) and SIC(ω').

Thus, diametrically opposite conclusions about the direction of change of the cost of living (and hence about changes in social welfare) result from the adoption of alternative social welfare functions. Of course, the same type of problem exists for an individual cost-of-living index: opposite conclusions about the direction of change of the cost of living can be obtained by the specification of alternative utility functions. But there is a crucial difference: assumptions about individual preferences are empirically testable, whereas assumptions about social welfare functions are not.

Jorgenson and Slesnick have jumped precariously from the familiar realm of positive econometrics to the ethereal realm of normative econometrics. The question is: who is to choose the social welfare function? I suspect that all would agree that neither a university economics professor nor a Bureau of Labor Statistics or Statistics Canada statistician should be saddled with this responsibility.

To be sure, Jorgenson and Slesnick are poignantly aware of this problem. It is for this reason that they painstakingly derive their social welfare function (using some powerful results of Roberts [1980]) from a set of axioms about the measurement of utility, interpersonal comparability, and social values (horizontal and vertical equity).[2] Nevertheless, this derivation – however ingenious – simply crystalizes the issue: none of these axioms is empirically refutable.[3] Moreover, the assumed social values, while couched euphemistically in terms of equity, are in fact strongly egalitarian: social optimality requires equal incomes for "scale-equivalent" households.

Even if we are prepared to accept a particular social welfare function – say, that proposed by Jorgenson and Slesnick – there is a second fundamental problem. While the assumption of individual utility maximization is almost tautological so long as we are willing to accept consistency of individual behaviour, there can be no comparable presumption that social welfare is maximized. Indeed, since maximization of the J-S social welfare function requires equal incomes for (scale-equivalent) households, it is evident that society does **not** maximize this social welfare function.

The implication of this problem for the interpretation of cost-of-living indexes is illustrated in Figure 3. Shown in this figure are two utility possibility frontiers, denoted $UPF(P^0,M)$ and $UPF(P^1,M)$, reflecting efficient utility profiles in two different periods with the same aggregate income, M, but different prices, P^0 and P^1. Maximum social welfare in period 0 is ω^0, as indicated by the tangency of $SIC(\omega^0)$ and $UPF(P^0,M)$.

The Pollak-Jorgenson-Slesnick social cost-of-living index would indicate an increase in the cost of living between periods 0 and 1, since aggregate income would have to be increased from M to $\bar{M}$ to attain social welfare ω^0 at prices P^1. That is, $UPF(P^1,\bar{M})$, which is tangent to the base-period social indifference curve, is everywhere above $UPF(P^1,M)$ and hence $\bar{M}$ must be above M. Thus, the social cost-of-living index calculation would suggest that welfare was higher in period 0 than in period 1. But, if society does not in fact maximize social welfare, then this need not be the case. For example, if the actual utility profiles were given by u^0 and u^1 in Figure 3, then welfare in period 1 would be higher than in period 0 for any welfare function incorporating the Pareto principle. Since society does not in fact maximize social welfare, the nexus between changes in the cost of living at given income levels and changes in welfare, stressed so much by Jorgenson and Slesnick, is broken.

The J-S model is illustrated, for the case of two households, in Figure 4. The translog individual utility functions generate concave utility possibility frontiers. (This is easily shown by substituting the translog indirect utility functions into equation (2) and applying the implicit function theorem to equation (3) to evaluate the rate of change u_i with respect to u_j). Along the ray where u_1 equals u_2, the slope of the utility possibility frontier is equal to the negative of the ratio of the scale factors, $-m_0(P,A_1)/m_0(P,A_2)$.[4]

The J-S social welfare function is a weighted average of individual-household indirect utility functions, with weights given by the household scale factors (normalized to sum to unity), plus a function of the scale factors and individual utilities that vanishes when all utilities are equal (a concave penalty function for income inequality).[5] The social indifference curves are therefore convex to the origin. Moreover, along the equal-utilities ray, the slopes of the indifference curves are equal to the negative of the ratio of the household scale factors, $-m_0(P,A_1)/m_0(P,A_2)$. (Were it not for the loss function for income inequality, each indifference curve would be linear with this slope.)

In Figure 4, $\mathrm{UPF}(P^0,M^0)$ is the utility possibility frontier in situation 0. Social optimality in the J-S model requires that individual utilities be equated, as reflected by the tangency of the social indifference curve, $\mathrm{SIC}(W(\overset{*}{u},P^0))$ and the utility possibility frontier at utility profile $\overset{*}{u}$. In reality, utilities are not equated and might be represented by a point such as u^0, with associated welfare $W(u^0,P^0)$.

Suppose that the utility possibility frontier shifts to $\mathrm{UPF}(P^1,M^1)$ in situation 1. The question that underlies the J-S social cost-of-living index is this: What level of aggregate expenditure allows society to obtain the **optimal** social welfare in situation 1, $W(\overset{*}{u},P^0)$, at situation-2 prices, P^2? The answer is M^1, the aggregate income that generates the utility possibility frontier that contains $\overset{*}{u}$ (the situation-1 optimal utility profile) – vis., $\mathrm{UPF}(P^1, M^1)$. The social welfare associated with the social indifference curve tangent to this utility possibility frontier, $W(\overset{*}{u},P^1)$, must be equal to the optimal social welfare in period 1, $W(\overset{*}{u},P^0)$. This is so even though the change in prices from P^0 to P^1 changes the weights in the social welfare function (and hence generates a new set of social indifference curves) because at equal incomes, such as at $\overset{*}{u}$, the social welfare is independent of the weights (and, of course, the penalty term for income inequality vanishes). Thus, in the J-S model, the social cost of living in situation 1 relative to situation 0 is equal to M^1/M_0, the ratio of aggregate income needed to obtain maximal social welfare in situation 0 at situation-1

Figure 4

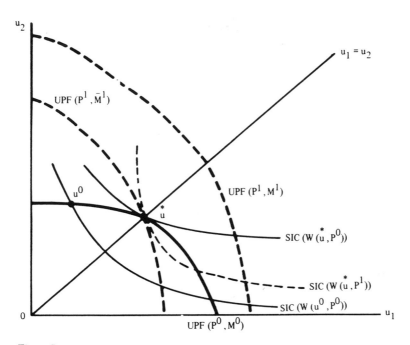

Figure 5

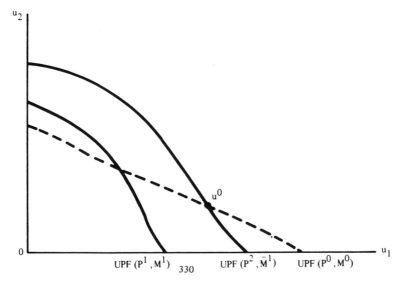

prices divided by actual aggregate income in situation 0. Thus, the J-S calculation compares only socially optimal points, even though society clearly does not optimize the J-S social welfare function.[6]

Thus, the J-S application of the theory of social cost-of-living indexes is plagued by both of the problems noted above. Cost-of-living comparisons are made along an equal-utilities ray (1) because of appealing but arbitrary normative assumptions and (2) despite the fact that actual utility profiles of society are not on this ray.

The question, of course, is this: What are the alternatives? Surely, as long as cost-of-living indexes are used for indexation of incomes of groups (such as unions, social security beneficiaries and welfare recipients), it makes little sense – at least from a theoretical point of view – to pretend that individual cost-of-living indexes are applicable for this purpose.[7] Three alternatives come to mind.

One alternative is to constrain individual utility functions sufficiently to ensure that the social cost-of-living index is independent of the form of the social welfare function. As Pollak [1981] showed, a sufficient condition for this independence is that aggregate demands be independent of the distribution of income. This is equivalent to individual preferences satisfying the condition of the Gorman polar form which is reflected in the following form of the expenditure function:

$$E^r(y,P) = \Lambda^r(P) + \pi(P)u_r \, ,$$

where Λ^r and π are positively linearly homogenous functions, and π is identical for all households.[8]

Equating the sum of these expenditure functions to aggregate expenditure (see equations (2)-(3)), we obtain the expression implicitly defining the utility possibility frontier:

$$\sum_r \Lambda^r(P) + \pi(P)\sum_r u_r = M.$$

It is apparent that the slope of the utility possibility frontier is everywhere equal to -1. Consequently the utility possibility frontiers generated by different price vectors and aggregate expenditures cannot cross, and the ranking of price-expenditure combinations is independent of the social welfare function. Thus, the cost-of-living calculation amounts simply to comparing utility possibility frontiers, with no meaningful reference to the social welfare function.

The trouble with this approach is that the assumption of linearity of the expenditure function in income is very strong. Moreover, empirical tests have rejected this property (see, for example, Pollak and Wales [1979]).

A second alternative is to eschew the social welfare approach altogether and to deal with cost-of-living changes for a group as a sort of statistical average. The idea is to calculate the change in the cost of living for an average (or representative) household with average income. The J-S model lends itself well to this approach, since the "average person" is well-defined in terms of the equivalence scales of their model. Indeed, where Jorgenson and Slesnick really end up, after invoking a medley of social choice results, **is** a statistical average. That is, the social cost-of-living function that they eventually employ is in fact the cost-of-living index for the average person with average income. Thus, one could just as well interpret the J-S calculations as those for the representative household, in which case one need not take into account their very strong assumptions about social values. The only assumptions that are relevant are those pertaining to individual utility functions, and these are empirically testable (at least in principle).

A third approach is to calculate a group cost-of-living index using utility possibility frontiers only – maintaining agnosticism about the existence as well as the form of a social welfare function. To illustrate, consider the base-period utility possibility frontier, $UPF(p^0, M^0)$, and base-period utility profile, u^0, in Figure 5. Suppose that the utility possibility frontier shifts to $UPF(p^1, M^1)$ in period 1. The proposed group cost-of-living index asks the following question: What aggregate income is required, at situation-1 prices, in order for the base-period utility profile u^0 to be feasible? In the picture, this expenditure level is $\overline{M}^1$, which, in combination with period-1 prices, P^1, generates a utility possibility frontier that contains u^0. Formally, this cost-of-living index, which we might call the Laspeyres group cost-of-living index, is given by

$$GCOL(u^0,M^0,P^1) = \sum_r E^r(u^0_r,P^1)/M^0.$$

(Note that this index implicity depends on base-period prices and income distribution, since u^0 can be replaced by $V^1(P^0/M^0_1),\ldots,V^R(P^0/M^0_R)$.)

We could similarly define a Paasche group cost-of-living index by adopting the period-1 utility profile u^1 as the base – i.e., by asking how much the base-period aggregate income would have to be adjusted in order to allow each household to obtain the utility level of situation 1.

The advantage of this approach is that it deals only with empirically refutable information about individual preferences, prices, and incomes; no assumptions or information about cardinal utility, interpersonal utility comparisons, or social values are required. Of course, the disadvantage is that the approach cannot be linked to social welfare. But there is no way to avoid this defect unless one is willing to make the strong assumptions that Jorgenson and Slesnick need for their approach. And even then, as noted above, the link between the index and social welfare is broken if society does not maximize welfare.

This approach to group cost-of-living indexes is analogous to the theory of Laspeyres and Paasche indexes as applied to individual consumers. The utility possibility frontier is analogous to the budget constraint and the utility profiles are analogous to consumption bundles.

If one were willing to assume that society does, indeed, maximize some type of social welfare function, but the index-number analyst does not know its form, then I suspect one could prove theorems relating the true but unknown group cost-of-living index to these Paasche and Laspeyres group indexes – that is, bounding theorems, analogous to those in Pollak [1971] and Diewert [1982]. (The non-linearity of the utility possibility frontiers would be problematic, but I don't believe that this would preclude getting some results.)

I do not mean to imply that any of these three alternatives is clearly preferable to the J-S approach. On the contrary, they are inchoate notions. The J-S contribution, on the other hand, is concrete. They might underestimate the effort required to implement – and, more important, to update – their model, but Jorgenson and Slesnick have gone far beyond all previous studies in making the social cost-of-living index a practical concept. Undoubtedly, much additional work needs to be done, but Jorgenson and Slesnick are to be commended for a promising and ambitious initiation of research on the empirical application of social cost-of-living indexes.

Footnotes

[1] The most immediate precursor is Jorgenson, Lau, and Stoker [1982]. For an earlier contribution to the empirical aggregation of demand systems (without the stringent requirements of exact aggregation in the sense of Gorman [1953]), see Berndt, Darrough, and Diewert [1977].

[2] Their specific functional form for the income-inequality part of the social welfare function – a mean-value function – additionally requires a strong additivity assumption.

[3] The authors seem to argue that cardinality is an implication of their aggregation conditions. This can't be correct. Linearity of the indirect utility function in expenditure (equation (2.6)) is sufficient, but not necessary, for aggregability. In short, the aggregation properties of their model are preserved by arbitrary monotone transformations of the indirect utility function.

[4] In the J-S model, the indirect utility functions are identical for all households except for household-specific scale factors, $m_0(P,A_r)$, that enter as deflators of money income. That is, the utility functions are identical in prices and scale-adjusted incomes.

[5] Since the scale factors are functions of prices, the J-S social welfare function is conditional on prices – directly as well as through the indirect effect of prices on individual utilities – i.e., the image is $W(u,P)$.

[6] The authors are not altogether clear about this fact. Equations (6.2) to (6.4), taken together, seem to imply that the social cost-of-living index is $W(u^0,P^0)/M^0$, the ratio of aggregate expenditure needed to obtain the **actual** social welfare in situation 0 at situation-1 prices, P^1, to the actual expenditure in situation 0. Such a comparison, however, would confound changes in the cost of living with changes in the distribution of income, since it would be comparing a social optimum in one situation with a suboptimal distribution of income in another. That this is not what the authors intend is indicated by the discussion following equations (6.2) to (6.4) and the discussion of Diagram 2, which compares only social optima.

[7] It should be noted that indexation of individual and group incomes is but one use of a consumer price index. Equally important is the use of such indexes to evaluate the macroeconomic performance of an economy. For this purpose, an index need not be based on the theory of individual utility maximization or social welfare. What is required is a reasonable measure of price inflation, and mechanistic indexes can serve tolerably well. For the purpose of evaluating macroeconomic performance, it would be useful to link a consumer price index to the welfare costs of inflation (i.e., to the induced-inefficiency costs of inflation attributable to the decreased information content of prices and the welfare losses associated with the attendant redistribution of income). But this welfare aspect of consumer price indices is quite different from the welfare issues addressed by Jorgenson and Slesnick (and others in session).

[8] The properties of the Gorman polar form are described in detail in Blackorby, Boyce, and Russell [1978] and Blackorby, Primont, and Russell [1978, ch. 8].

References

Berndt, E.R., M.N. Darrough and W.E. Diewert [1977], "Flexible Functional Forms and Expenditure Distributions: An Application to Canadian Consumer Demand Functions", *International Economic Review* 18, pp.651-676.

Blackorby, C., R. Boyce and R.R. Russell]1978], "Estimation of Demand Systems Generated by the Gorman Polar Form: A Generalization of the S-Branch Utility Tree", *Econometrica* 46, pp.345-364.

_____ D. Primont and R.R. Russell [1978], *Duality Separability, and Functional Structure: Theory and Economic Applications*, New York: North-Holland.

Diewert, W.E. [1983], "The Theory of the Cost-of-Living Index and the Measurment of Welfare Change", *Price Level Measurement: Proceedings From a Conference Sponsored by Statistics Canada.*

Gorman, W.M. [1953], "Community Preference Fields", *Econometrica* 21, pp.63-80.

Jorgenson, D.W., L.J. Lau and T.M. Stoker [1980], "Welfare Comparison Under Exact Aggregation", *American Economic Review* 70, pp.268-272.

Pollak, R.A. [1971], "The Theory of the Cost-of-Living Index", Research Discussion Paper II, Office of Prices and Living Conditions, U.S. Bureau of Labor Statistics, Washington, D.C.

_____ [1981], "The Social Cost-of-Living Index", *Journal of Public Economics* 15, pp.311-336.

_____ and T.J. Wales [1979], "Welfare Comparisons and Equivalent Scales", *American Economic Review* 69, pp.216-21

Roberts, K.W.S. [1980], "Interpersonal Comparability and Social Choice Theory", *Review of Economic Studies* 47, pp.421-439.

PRICE LEVEL MEASUREMENT – W.E. Diewert (Editor)
Canadian Government Publishing Centre /
Elsevier Science Publishers B.V. (North-Holland)
© Minister of Supply and Services Canada, 1990

LEISURE TIME AND THE MEASUREMENT OF ECONOMIC WELFARE

W. Craig Riddell
Department of Economics
University of British Columbia

SUMMARY

Most measures of economic welfare and the cost of living are based on consumer expenditure on goods and services. Such measures omit an important commodity – leisure time – which is purchased implicitly by not working. This paper discusses the issues involved in incorporating leisure time in measures of the standard of living, and empirically implements the theoretical discussion by calculating some measures of economic welfare for Canada which incorporate leisure.

In order to calculate true measures of economic welfare, knowledge of the preferences of individual consumers or households is required. Pencavel [1977] provides measures of welfare for the U.S. using the estimated parameters of the Stone-Geary utility function. The advantage of this procedure is that it allows for substitution among goods and leisure in response to changes in relative prices. The disadvantage is that it assumes a specific functional form for the consumer's utility function.

This paper employs index number theory to obtain measures of economic welfare in the joint commodity demand-labour supply context. Thus the measures obtained are not based on any assumed functional form for the utility function. However, the index numbers do not fully allow for substitution in response to changes in relative prices and are thus only approximations to the true measures of changes in the standard of living.

A related issue addressed in this paper is the interpretation of published real wage or real earnings indexes, obtained by deflating nominal wages or earnings by a price index.

While these are related to economic welfare, neither is a legitimate measure of the standard of living. Earnings may increase because of an increase in hours worked at existing hourly wage rates or because of an increase in the hourly wage with no change in hours of work. The two have very different implications for welfare change. Similarly real wage indexes are not satisfactory measures of welfare because they do not allow for leisure time. However, real wage and real earnings indexes which are legitimate measures of the standard of living can be constructed; the details are given in the paper.

1. Introduction

One of the most important commodities (indeed, probably **the** most important commodity omitted from measures of economic welfare and the cost of living is leisure time. There is, of course, a good reason for this omission. Leisure time is purchased implicitly (by not working) rather than explicitly, and the price associated with an hour of leisure is also implicit. Nonetheless there are also good reasons for attempting to incorporate leisure into measures of the cost of living, real wages and real income. Observed changes in earnings may have come about because of changes in wage rates or changes in hours worked or various combinations of the two. Yet for measures of welfare the composition of the change in earnings is important. Further, observed changes in earnings include hours of work decisions which themselves reflect adjustments to changes in wage rates and these underlying time allocation choices should be incorporated in measures of welfare.

This paper will discuss the issues involved in incorporating leisure time in measures of the cost of living, real wages and real income and will empirically implement the discussion by deriving some index numbers which incorporate leisure.

The next section of the paper briefly summarizes the theory of index numbers of real income in the context of the standard model of consumer behaviour. This material is well-known, and is included here because it provides useful background for the theory of index numbers of real wages or real income in the context of the income-leisure choice model of consumer-worker behaviour. Section 3 contains this discussion of the theory of index numbers of real income in the joint consumption and labour supply case. This material is not well-known, the main contributions having been made quite recently (specifically, Pencavel [1977, 1979a], Lloyd [1979], Cleeton [1982]).

The CPI is widely used as a measure of the cost of living. It is a relatively straightforward matter to incorporate leisure time into measures of the cost of living, if this is desired. The Family Expenditure Survey, upon which the weights of CPI are based, could be expanded to include questions about hours of work and hourly wage rates. Hours of leisure could be treated in the same fashion as an explicitly purchased commodity, the price of leisure being the after-tax hourly wage rate. The main complications are two: (1) the marginal price (the price of an additional hour of leisure) may not equal the average price, because of tax considerations (with a progressive income tax the marginal after-tax wage may differ from the average after-tax wage) and possibly because of hours of work constraints (increasing leisure time may involve switching jobs, at least in the short run; reducing leisure may involve a second job with a different wage than the primary job), and (2) the price of leisure differs across individuals (because of differences in after-tax wage rates) to a much greater extent than do the prices of other commodities. Nonetheless, these complicating factors are not particularly severe.

While some discussion of measures of the cost of living will be provided, the main focus of the paper is on measures of the standard of living, real income, or economic welfare (these three terms will be used synonomously). The CPI is frequently used to deflate money incomes or money wages, producing measures of real incomes or real wages. Are these legitimate measures of the standard of living? Real income measures are typically based on the standard theory of consumer behaviour in which the consumer's money income is taken to be exogenous. However, once the individual's time constraint and the concomitant labour supply decision are recognized, the income available for expenditure on goods is chosen by the consumer-worker. In this setting, most measures of real income are not true measures of the standard of living. Section 3 of the paper discusses two concepts that can be used for this purpose: real "non-labour income" and real "full income" (defined later).

Similarly, deflating nominal wages by the CPI does not produce a true measure of economic welfare. Real wage indices that have this property can, however, be constructed and these are discussed in Section 3. Because of the importance of labour income to most individuals, such real wage measures are likely to be of more interest than real non-labour income measures.

Pencavel [1977] appears to have been the first author to attempt to construct index numbers of real wages and real non-labour income that are true measures of the standard of living. His procedure employed the estimated parameters of a direct utility function and was based on the income-leisure choice model. (The estimates were obtained from Abbott and Ashenfelter's [1976] study of commodity demand and labour supply.) The advantage of this procedure is that it allows for substitution among goods and leisure in response to changes in relative prices. The disadvantage is that it assumes a specific functional form for the consumer's utility function (the Stone-Geary utility function in Pencavel's study).

This paper employs index number theory to obtain measures of real income and real wages in the joint commodity demand-labour supply context. Thus the measures obtained are not based on any assumed functional form for the utility function. However, the index numbers do not fully allow for substitution in response to changes in relative prices and are thus only approximations to the true measures of changes in the standard of living.

2. Index Numbers of Real Income and The Cost of Living

The purpose of this section is to summarize some known results on index numbers which are needed later in the paper. Proofs of these results together with references to the previous literature may be found in the survey paper of Diewert [1981].

Measures of real income and the cost of living are usually based on the standard model of consumer behaviour. Let $U(x)$ be the individual's ordinal utility function where $x = (x^1...x^n)$ is a 1 by n vector and x^i is the quantity of the ith good. Let $p = (p^1...p^n)$ be the vector of prices of the n goods and y be the consumer's income. If p and y are exogenous, the consumer will choose quantities consumed to

$$\underset{(x)}{\text{Max}} \ U(x) \qquad\qquad\qquad (A)$$

$$\text{subject to } p \cdot x = y$$

where $p \cdot x$ is the inner product of the vectors p and x.

Let $x^*(p,y)$ be the solution to (A); this is the set of commodity demand functions. Substituting into $U(x)$ gives the indirect utility function

$$V(p,y) = U(x^*(p,y)) \qquad\qquad\qquad (1)$$

which shows the maximum utility as a function of the exogenous variables prices and income. Inverting (1) gives the expenditure function $e(p,\bar{U})$ which solves

$$\underset{(x)}{\text{Min}} \ p \cdot x \qquad\qquad\qquad (B)$$

$$\text{subject to } U = \bar{U}$$

That is, $e(p,\bar{U})$ shows the minimum expenditure needed to achieve utility $\bar{U}$ at prices p.

Now suppose that prices and income vary over time (or across regions or countries) and we wish to construct measures of the consumer's welfare or standard of living. Let (p_0,x_0,y_0) and (p_1,x_1,y_1) be the observed prices, quantities consumed and expenditure in periods 0 and 1 respectively and let $Q(p_0,p_1,x_0,x_1)$ be an index number of real income. Such an index should have the property

$$Q(p_0,p_1,x_0,x_1) > 1 \text{ if and only if } V(p_1,y_1) > V(p_0,y_0) \tag{P1}$$

That is, the index "must itself be a cardinal indicator of ordinal utility" (Samuelson and Swamy, [1974, p.568]). Two such measures are

$$Q_{LA} \equiv e(p_0,U(x_1))/e(p_0,U(x_0)) \tag{2}$$

$$Q_{PA} \equiv e(p_1,U(x_1))/e(p_1,U(x_0)) \tag{3}$$

Q_{LA} is the Laspeyres-Allen index of real income and Q_{PA} is the Paasche-Allen index of real income. These true, or constant-utility, index numbers of real income are, of course, unobservable. However, $p_0 \cdot x_1 \geq e(p_0,U(x_1))$ and $p_1 \cdot x_0 \geq e(p_1,U(x_0))$ so that upper and lower bounds can be obtained. Thus

$$Q_{LA} \leq \frac{p_0 \cdot x_1}{p_0 \cdot x_0} \equiv Q_L \tag{4}$$

$$Q_{PA} \geq \frac{p_1 \cdot x_1}{p_1 \cdot x_0} \equiv Q_P \tag{5}$$

where Q_L is the Laspeyres quantity index and Q_P is the Paasche quantity index.[1]

Another constant-utility index number of real income is the Malmquist index. The Laspeyres-Malmquist index Q_{LM} and Paasche-Malmquist index Q_{PM} are given by

$$Q_{LM} \equiv D(U(x_0), x_1) \tag{6}$$

$$Q_{PM} \equiv 1/D(U(x_1), x_0) \tag{7}$$

where $D(U(x), \bar{x})$ is the deflation function defined by

$$D(U,\bar{x}) \equiv \underset{k>0}{\text{Max}} \left\{ k : U(\bar{x}/k) \geq U \right\} \tag{8}$$

One advantage of the Malmquist indices is that both upper and lower bounds exist for each:

$$\text{Min}\left\{x_1^i/x_0^i; \; i=1,....n\right\} \leq Q_{LM} \leq Q_L \tag{9}$$

and

$$Q_P \leq Q_{PM} \leq \underset{i}{\text{Max}}\left\{x_1^i/x_0^i; \; i=1....n\right\} \tag{10}$$

Two other quantity indices are Fisher's ideal quantity index Q_F and the Tornqvist or translog quantity index Q_T. These are defined by

$$Q_F \equiv (Q_L \cdot Q_P)^{1/2} \tag{11}$$

$$Q_T \equiv \prod_{i=1}^{n} (x_1^i/x_0^i)^{1/2(s_0^i + s_1^i)} \tag{12}$$

where s_t^i is commodity i's share of total expenditure in period t.

$$s_t^i \equiv p_t^i x_t^i/p_t \cdot x_t \tag{13}$$

Turning now to index numbers of the cost of living, the Laspeyres-Konüs P_{LK} and Paasche-Konüs P_{PK} constant-utility index numbers are

$$P_{LK} \equiv e(p_1, U(x_0))/e(p_0, U(x_0)) \tag{14}$$

$$P_{PK} \equiv e(p_1, U(x_1))/e(p_0, U(x_1)) \tag{15}$$

These are bounded by

$$\underset{i}{\text{Min}}\left\{p_1^i/p_0^i; \; i=1,...,n\right\} \leq P_{LK} \leq P_L \tag{16}$$

$$P_P \le P_{PK} \le \underset{i}{\text{Max}} \left\{ p_1^i/p_0^i; \ i = 1,...,n \right\} \tag{17}$$

where P_L and P_P are the Laspeyres and Paasche price indices defined by

$$P_L \equiv p_1 \cdot x_0/p_0 \cdot x_0 \tag{18}$$

$$P_P \equiv p_1 \cdot x_1/p_0 \cdot x_1 \tag{19}$$

The CPI is, of course, a Laspeyres price index.

Two additional price indices are Fisher's ideal price index P_F and the Tornqvist or translog price index P_T:

$$P_F \equiv (P_P P_L)^{1/2} \tag{20}$$

$$P_T \equiv \prod_{i=1}^{n} (p_1^i/p_0^i)^{1/2(s_0^i + s_1^i)} \tag{21}$$

The standard model of consumer behaviour, and thus the index numbers based on that model, have a number of important limitations. Most important for present purposes is that while prices are typically exogenous to the individual consumer or household, income is not. Income and expenditure on goods are chosen by individuals. Three key aspects of this choice are the following. First, individuals facing exogenous hourly wages can determine income by choosing the number of hours per week and year to work. The ultimate constraint the individual faces is the time, not income, constraint. Second, individuals can influence the (hourly) wage rate they receive by investing in human capital (education, training, migration, etc.). Third, expenditure on goods need not equal income each period.

The latter two aspects influencing observed income and expenditure require an intertemporal choice model. Incorporating such life cycle aspects is outside of the objectives of this paper. The first aspect involves incorporating leisure as a commodity chosen by the individual. The most basic model which accomplishes this is the income-leisure choice model, and it is to the measurement of the standard of living in the context of this model that we now turn.

3. Income-Leisure Choice and Index Numbers of Real Income

The individual now chooses the vector of goods x and hours of leisure 1 to

$$\underset{(x,l)}{\text{Max}} \quad U(x,l) \qquad\qquad\qquad \text{(B)}$$

subject to $\qquad\qquad$ p· x = wh + m

and $\qquad\qquad$ h + 1 = T.

Here h is hours worked, T is the total time available for the two activities, work and leisure, m is non-labour income, and w is the after-tax hourly wage rate. The variables T, m and w are taken to be exogenous, although as mentioned above both m and w would become choice variables in an intertemporal model. The price of leisure is the wage rate, as is made clear by writing the constraint as follows:

$$p·x + wl = wT + m = Y \qquad\qquad\qquad \text{(22)}$$

where Y = wT + m is called "full income" and is exogenous to the consumer.[2] The constraint (22) is equivalent to that in the maximization problem (A); the left-hand side is total expenditure on goods including leisure and the right-hand side is exogenous income.

Measures of welfare can be based on any of the variables which are exogenous to the decision maker. In the standard theory of consumer behaviour, income and prices are treated as exogenous. Either could be used to construct constant utility index numbers; income is usually chosen because comparisons based on income are more meaningful than those based on any particular price. In the case of joint consumption and labour supply behaviour, the exogenous variables are prices p, the wage rate w, non-labour income m and total time available. Again, comparisons based on the price of a particular market good are unlikely to be very meaningful. Comparisons based on the price of the commodity leisure are, however, meaningful. Thus index numbers of real wages are potentially useful as measures of welfare change. Similarly, index numbers of real non-labour income may be employed.

Finally, index numbers of time available could also be constructed, although I am not aware of any such measures having been calculated. Presumably this reflects the fact that total time available T is not only exogenous but also fixed. Index numbers based on T would involve the conceptual experiment of altering the time available to the consumer-worker.

Attention can thus be focused on two welfare measures – index numbers of real wages and real non-labour income. As before, these are defined in terms of the value of the exogenous variables in the current period that would provide the consumer-worker with the same utility as in the base period.

With joint commodity demand and labour supply decisions, the indirect utility function becomes $V(p,w,m)$ and the expenditure function becomes $c(p,w,U)$. (The argument T has been suppressed because T is fixed.) A constant-utility index of real non-labour income is defined analogously to that of real income in the previous section. Define m_1^* as the minimum non-labour income required in period 1 to attain base period utility $V(p_0,w_0,m_0)$; that is;

$$m_1^* = c(p_1,w_1,U(x_0,l_0))   \qquad (23)$$

Thus

$$V(p_1,w_1,m_1^*) = V(p_0,w_0,m_0)   \qquad (24)$$

and m_t/m_t^* is a constant-utility index number of real non-labour income. m_1/m_1^* corresponds to the Laspeyres-Allen index Q_{LA} defined earlier in the context of the standard theory of consumer behaviour. Similarly there is a Paasche-Allen index number of real non-labour income which uses period 1 prices and wage rate as the base. These two measures are given by

$$M_{LA} \equiv c(p_0,w_0,U(x_1,l_1))/c(p_0,w_0,U(x_0,l_0))   \qquad (25)$$

$$M_{PA} \equiv c(p_1,w_1,U(x_1,l_1))/c(p_1,w_1,U(x_0,l_0))   \qquad (26)$$

Note that M_{LA} and M_{PA} may be negative. Indeed, with wage rates rising over time we would expect $c(p_0, w_0, U(x_1, l_1))$ to become large relative to $c(p_0, w_0, U(x_0, l_0))$ so that M_{LA} will become large and we would expect $c(p_1, w_1, U(x_0, l_0))$ to possibly become negative, making M_{PA} negative.

Just as the Laspeyres-Allen index of real income Q_{LA} is bounded from above by the Laspeyres quantity index Q_L and the Paasche-Allen index of real income Q_{PA} is bounded from below by the Paasche quantity Q_P, the index numbers of real non-labour income are bounded by corresponding quantity indices M_L and M_P defined by

$$M_L \equiv (p_0 \cdot x_1 - w_0 h_1)/(p_0 \cdot x_0 - w_0 h_0) \tag{27}$$

$$M_P \equiv (p_1 \cdot x_1 - w_1 h_1)/(p_1 \cdot x_0 - w_1 h_0) \tag{28}$$

Theorem 1 (Cleeton [1982, p.224]):

$$M_{LA} \leq M_L \tag{29}$$

$$M_{PA} \geq M_P \tag{30}$$

Proof: From utility maximization, the denominators of M_{LA} and M_L are equal; that is,

$$p_0 \cdot x_0 - w_0 h_0 = m_0 = c(p_0, w_0, U(x_0, l_0)) \tag{31}$$

Comparing the numerators,

$$c(p_0, w_0, U(x_1, l_1)) = \mathrm{Min} \left\{ (p_0 \cdot x - w_0 l) : U(x, l) \geq U(x_1, l_1) \right\}$$

$$\leq p_0 x_1 - w_0 h_1 \tag{32}$$

since (x_1, l_1) is feasible for the minimization problem but not necessarily optimal. The proof of (30) is similar.

Index numbers could be defined in terms of full income Y rather than non-labour income m. Let f(p,w,U) be the minimum full income expenditure function; that is,

$$f(p,w,\bar{U}) \equiv \underset{(x,l)}{\mathrm{Min}} \left\{ (p \cdot x + wl) : U(x,l) \geq \bar{U} \right\} \tag{33}$$

The Allen index numbers of real full income are

$$Y_{LA} \equiv f(p_0,w_0,U(x_1,l_1))/f(p_0,w_0,U(x_0,l_0)) \tag{34}$$

$$Y_{PA} \equiv f(p_1,w_1,U(x_1,l_1))/f(p_1,w_1,U(x_0,l_0)) \tag{35}$$

and the Laspeyres and Paasche quantity index numbers are

$$Y_L \equiv (p_0 \cdot x_1 + w_0 l_1)/(p_0 \cdot x_0 + w_0 l_0) \tag{36}$$

$$Y_P \equiv (p_1 \cdot x_1 + w_1 l_1)/(p_1 \cdot x_0 + w_1 l_0) \tag{37}$$

These quantity index numbers provide one-sided bounds for the corresponding constant-utility index numbers of full income.

Theorem 2:

$$Y_{LA} \leq Y_L \tag{38}$$

$$Y_{PA} \geq Y_P \tag{39}$$

Proof: Similar to Theorem 1.

Index numbers of real non-labour income are more intuitively appealing than those of real full income as few, if any, individuals think in terms of their full income (i.e., the sum of non-labour income and what labour income would be if no leisure time were consumed). This suggests using the quantity index numbers M_L and M_p in lieu of Y_L and

Y_p. One point, however, which may be considered an argument in favour of the full income quantity indices is that they provide tighter bounds than their non-labour income counterparts.

Theorem 3:

$$Y_L - Y_{LA} \leq M_L - M_{LA} \tag{40}$$

$$Y_{PA} - Y_P \geq M_{PA} - M_P \tag{41}$$

Proof: From the definitions of the minimum non-labour income and minimum full income functions

$$c(p_0,w_0,U(x_1,l_1)) + w_0 T = f(p_0,w_0,U(x_1,l_1)) \tag{42}$$

Using the definitions (25) and (27), utility maximization implies that

$$M_L - M_{LA} = (p_0 \cdot x_1 - w_0 h_1 - c(p_0,w_0,U(x_1,l_1)))/m_0 \tag{43}$$

and, using (34) and (36),

$$Y_L - Y_{LA} = (p_0 \cdot x_1 + w_0 l_1 - f(p_0,w_0,U(x_1,l_1)))/Y_0 \tag{44}$$

Now note that (42) implies that the numerator of the right-hand side of (43) equals the numerator of the right-hand side of (44). However, $m_0 \leq Y_0$ which establishes (40). The proof of (41) is similar.

While Theorem 3 provides some justification for index numbers of real full income, two factors make use of such measures unattractive. One was mentioned above – few individuals think in terms of full income, so the measure does not have broad intuitive appeal. Second, the indices Y_{LA} and Y_{PA} contain an element of arbitrariness in the measurement of leisure l. Do we take T = 24 hours per day or T = 17, allowing seven hours for activities which are neither leisure nor paid work (mainly sleeping and eating). A variety of alternative assumptions have been made in empirical work; see the discussion in Pencavel [1979b] or Usher [1980, pp.135-47]. This element of arbitrariness is not present in the non-labour income indices M_{LA} and M_{PA}.

Thus measures of both real non-labour income and real full income have some unattractive features. Because of the importance of labour income to most individuals, an appealing alternative is a real wage index. Indeed, real wages are frequently used along with real income as indicators of economic welfare.

Pencavel [1977] appears to have been the first author to construct index numbers of real wages that are true measures of the standard of living. The theory of real wage indices for this purpose was subsequently extended by Cleeton [1982]. To begin, define the minimum wage function as

$$w(p,m,\bar{U}) \equiv \underset{(x,l)}{\text{Min}} \left\{ (p \cdot x - m)/h \; ; \; U(x,l) > \bar{U}, h + l = T \right\} \tag{45}$$

Thus $w(p,m,\bar{U})$ shows the minimum hourly wage rate required to attain utility $\bar{U}$ at prices p and non-labour income m. $w(p,m,\bar{U})$ is not defined if the individual's non-labour income is sufficiently great that $\bar{U}$ can be attained without working; i.e., if $U(x^*,T) > \bar{U}$ where $p \cdot x^* \leq m$.

The Allen index numbers of real wages can be defined in an analogous fashion to real income indices:

$$W_{LA} \equiv w(p_0,m_0,U(x_1,l_1))/w(p_0,m_0,U(x_0,l_0)) \tag{46}$$

$$W_{PA} \equiv w(p_1,m_1,U(x_1,l_1))/w(p_1,m_1,U(x_0,l_0)) \tag{47}$$

These true cost-of-living index numbers of real wages can be approximated by their Laspeyres and Paasche quantity index counterparts:

$$W_L \equiv [(p_0 \cdot x_1 - m_0)/h_1]/w_0 \tag{48}$$

$$W_P \equiv w_1/[p_1 \cdot x_0 - m_1)/h_0] \tag{49}$$

Theorem 4 (Cleeton [1982, p.223]):

$$W_{LA} \leq W_L \tag{50}$$

$$W_{PA} \geq W_P \tag{51}$$

Proof: From utility maximization,

$$w_0 = w(p_0, m_0, U(x_0, l_0)) \tag{52}$$

Comparing the numerators of (46) and (48):

$$W(p_0, m_0, U(x_1, l_1)) = \text{Min} \left\{ (p_0 \cdot x - m_0)/h : U(x, l) \geq U(x_1, l_1) ; h + l = T \right\}$$

$$\leq (p_0 \cdot x_1 - m_0)/w_0 \tag{53}$$

since (x_1, l_1) is feasible for the minimization problem but not necessarily optimal. The proof of (51) is similar.

Note that neither of the indices (46) or (47) or their observable counterparts (48) and (49) are ''real wage indices'' as the term is generally used. Deflating the money wage by the CPI does not provide a true measure of the standard of living for two reasons: (1) the contribution of non-labour income to welfare is ignored and (2) the contribution of leisure time to welfare is ignored.

The literature on index numbers of real wages and real income in the context of the income-leisure choice model has confined its attention to Allen-type indices of economic welfare. We now turn to the nature and properties of Malmquist-type indices in this setting.

With joint consumption and labour supply decisions, the Laspeyres-Malmquist index R_{LM} and Paasche-Malmquist index R_{PM} can be defined as follows:

$$R_{LM} \equiv D(U(x_0, l_0), (x_1, l_1)) \tag{54}$$

$$R_{PM} \equiv 1/D(U(x_1, l_1), (x_0, l_0)) \tag{55}$$

where the deflation function D is defined analogously to (8):

$$D(U, (\bar{x}, \bar{l})) \equiv \text{Max} \left\{ k : U(\bar{x}/k, \bar{l}/k) \geq U \right\} \tag{56}$$

Note that the Malmquist index of real income is not defined in terms of the amount of any particular exogenous variable required to attain the reference level of economic welfare. Thus there is no need to consider Malmquist indices of non-labour income, full income, or the wage rate. Nonetheless, any one of these three variables may be used to approximate a Malmquist true welfare index. For this reason it is useful to state results on the bounds of the Malmquist indices in two stages.

Theorem 5 (Diewert [1981, p. 175]):

$$\underset{i}{\text{Min}} \left\{ x_1^i / x_0^i; \ i = 1, \dots. n + 1 \right\} < R_{LM} < \underset{i}{\text{Max}} \left\{ x_1^i / x_0^i; \ i = 1, \dots. n + 1 \right\} \tag{57}$$

$$\underset{i}{\text{Min}} \left\{ x_1^i / x_0^i; \ i = 1, \dots. n + 1 \right\} \leq R_{PM} \leq \underset{i}{\text{Max}} \left\{ x_1^i / x_0^i; \ i = 1, \dots. n + 1 \right\} \tag{58}$$

$$\text{where} \quad x_t^{n+1} \equiv l_t \tag{59}$$

Proof: The theorem is a straightforward generalization of Theorem 13 in Diewert [1981]. Adding the labour supply or leisure choice dimension does not alter the structure of the proof.

The bounds given in (57) and (58) may not be particularly tight ones. In the standard consumer theory (income treated as exogenous) the Malmquist indices are also bounded by the Laspeyres and Paasche quantity indices, and these provide a tighter upper bound for Q_{LM} and a tighter lower bound for Q_{PM} (recall equations (9) and (10)). In the income-leisure choice model we have discussed three alternative sets of Laspeyres and Paasche quantity indices (defined in terms of non-labour income, full income and the wage rate) and we now wish to investigate the relationship between these and the Malmquist indices R_{LM} and R_{PM}.

Theorem 6:

$$R_{LM} \leq Y_L \leq \underset{i}{Max} \left\{ x_1^i / x_0^i; \, i = 1, \dots n+1 \right\} \tag{60}$$

$$\underset{i}{Min} \left\{ x_1^i / x_0^i; \, i = 1, \dots n+1 \right\} \leq Y_P \leq R_{PM} \tag{61}$$

Proof: (a) Proof that $Y_L \leq \underset{i}{Max} \left\{ x_1^i / x_0^i \, ; \, i = 1, \dots n+1 \right\}$

$$Y_L = \sum_{i=1}^{n+1} p_0^i x_1^i / (p_0 \cdot x_0 + w_0 l_0) \qquad \text{from (36) and (59)}$$

$$= \sum_{i=1}^{n} s_1^i x_1^i / x_0^i \qquad \text{from (13)}$$

$$\leq \underset{i}{Max} \left\{ x_1^i / x_0^i; \, i = 1, \dots n+1 \right\}$$

since a share-weighted average must be less than the maximum value entering into the average.

(b) Proof that $R_{LM} \leq Y_L$

Let k^* be the solution to (54) and (56). That is,

$$R_{LM} \equiv \underset{k>0}{\text{Max}} \left\{ k : U(x_1/k, l_1/k) \geq U(x_0,l_0) \right\}$$

$$= k^*$$

Now $p_0 \cdot x_0 + w_0 l_0 = f(p_0, w_0, U(x_0,l_0))$

$$\equiv \underset{x,l}{\text{Min}} \left\{ p_0 \cdot x + w_0 l : U(x,l) \geq U(x_0,l_0) \right\}$$

$$\leq p_0 \cdot x_1/k^* + w_0 l_1/k^*$$

since $(x_1/k^*, l_1/k^*)$ is feasible for the minimization problem.

Thus

$$k^* \leq (p_0 \cdot x_1 + w_0 l_1)/(p_0 \cdot x_0 + w_0 l_0) \equiv Y_L$$

The proof of (61) is similar

Theorem 6 provides additional justification for using index numbers of real full income rather than non-labour income. In particular, it is not true that

$$M_L \leq \underset{i}{\text{Max}} \left\{ x_1^i/x_0^i; \ i = 1,\ldots n+1 \right\} .$$

The Tornqvist index generalizes easily to the income-leisure choice model.

$$Y_T \equiv \prod_{i=1}^{n+1} (x_1^i/x_0^i)^{1/2(s_0^i + s_1^i)} \tag{62}$$

where the shares are defined in terms of full income

$$Y = p \cdot x + wl = \sum_{i=1}^{n+1} p_i x_i \tag{63}$$

The Fisher ideal index can be defined in terms of full income

$$Y_F \equiv (Y_L \, Y_P)^{1/2} \tag{64}$$

However it is not possible to define the equivalent Fisher index in terms of non-labour income because M_L and M_P may be negative. The real wage indices do not have this property, however, so that we could define

$$W_F \equiv (W_L \, W_P)^{1/2} \tag{65}$$

Real wage index numbers are one alternative to index numbers based on non-labour income or full income. Another alternative, suggested recently by Diewert [1983] is an index based on consumption expenditure in the base period. Specifically, Diewert proposes the measures

$$R_{LDM} \equiv F(U(x_0,l_0),(x_1,l_1)) \tag{66}$$

$$R_{PDM} \equiv 1/F(U(x_1,l_1),(x_0,l_0)) \tag{67}$$

where the deflation function is defined in terms of consumption of commodities other than leisure:

$$F(U,(\bar{x},\bar{l})) \equiv \text{Max} \left\{ k: U(\bar{x}/k,\bar{l}) \geq U \right\} \tag{68}$$

This deflation function can be compared to the function D in (56) which deflates the vector $(\bar{x}, \bar{l})$ rather than the goods consumption component $\bar{x}$. Geometrically, the Laspeyres-Diewert-Malmquist and Paasche-Diewert-Malmquist index numbers R_{LDM} and R_{PDM} deflate goods consumption along a vertical ray based on base period leisure time whereas the index numbers R_{LM} and R_{PM} deflate goods consumption and leisure time along a ray through the origin.

Observable approximations to R_{LDM} and R_{PDM} are the Laspeyres and Paasche consumption quantity index numbers E_L and E_P:

$$E_L \equiv \frac{p_0 \cdot x_1}{p_0 \cdot x_0 + w_0(h_1 - h_0)} \qquad (69)$$

$$E_P \equiv \frac{p_1 \cdot x_1 + w_1(h_0 - h_1)}{p_1 \cdot x_0} \qquad (70)$$

Theorem 7 (Diewert [1982, p.47]):

$$R_{LDM} \leq E_L \qquad (71)$$

$$R_{PDM} \geq E_P \qquad (72)$$

Proof: Similar to Theorem 1.

The index numbers E_L and E_P have two advantages over other measures which also incorporate leisure time: (1) they are based on an intuitively appealing quantity, expenditure on goods (2) they do not require an arbitrary assignment of total time available.

Index numbers of the cost of living which incorporate leisure as a commodity can also be defined in terms of the three exogenous variables non-labour income, full income and the after-tax hourly wage rate. These are discussed later. We now turn to some empirical measures of real income and real wages.

4. Index Numbers of Real Income and Real Wages: Canada, 1949-80

Several measures of real income and real wages are presented in this section. The data are annual, and cover the period 1949-80. The base year is 1949. With the exception of the hours worked series the data are from the National Accounts. The commodity breakdown is as follows:

1. Food, beverages and tobacco
2. Clothing and footwear
3. Rent, fuel and power

4. Furniture, furnishings, household equipment and operating costs

5. Medical care and health services

6. Transportation and communications (including net foreign expenditures abroad)

7. Recreation, entertainment, education and cultural services

8. Personal goods and services.

Thus there are eight commodities plus leisure. Net foreign expenditures abroad are included in transportation and communications because these are primarily travel expenditures.

At this level of aggregation, the commodities are themselves quantity indices. Further research with substantially more disaggregate data is planned; however, even at this aggregate level the main consequences of incorporating leisure in measures of economic welfare should emerge.

The hours worked data are from the Labour Force Survey. To be as consistent as possible with the use of national accounts data on income and expenditures, the broadest available measure of hours worked is used. This is actual average hours worked per week in all jobs, the average calculated excluding persons who were not at work during the reference week. The latter ensures that annual hours take account of vacations, etc. Using average hours based on the total employed during the reference week (i.e., including persons who were not at work during the reference week) would overstate annual hours worked unless some alternative correction for vacation time were made. This factor is important because in the period under consideration the increase in leisure time has come about not only from a reduction in hours worked per week but also from a reduction in days worked per year.

Labour income, non-labour income and total expenditure on goods and services (other than leisure) are taken from the income and outlay account of the National Income and Expenditure Accounts. Current transfers to government and other sectors (primarily income taxes) is subtracted from labour income to give after-tax labour income. Dividing this by annual hours worked gives the after-tax wage rate. Non-labour income is net of savings, so that this is non-labour income actually spent. This convention ensures that the budget constraint is met; i.e., that after-tax labour income plus non-labour income sum to total expenditure on goods and services.

Table 1 presents the quantity indices discussed in Section 2 of the paper; i.e., those based on the standard theory of consumer behaviour. These employ only the personal expenditure data from the national accounts; i.e., they ensure that the budget constraint is satisfied by assuming that income equals total personal expenditure.

The four indices Q_L, Q_P, Q_F and Q_T lead to similar conclusions. Real income approximately doubled from 1949 to 1964 and from 1949 to 1976. Since then real income declined slightly.

The table also confirms the expectation that the Min and Max bounds are likely to be too broad to be useful as measures of real income.

Tables 2 and 3 contain index numbers which incorporate income-leisure choice. The various measures based on full income are shown in Table 2. For the full income measures it was assumed that 16 hours per day are available for work or leisure activities.

The index numbers Y_L, Y_P, Y_F and Y_T are very similar in magnitude (indeed, Y_F and Y_T are almost identical throughout the period). They are also consistently lower than their counterparts in Table 1. Real full income rises by, at most, 50 percent from 1949 to 1976, whereas the indices in Table 1 suggest that real income more than quadrupled over this period. The reason for these differences is that the quantity of leisure consumed has not increased at the same rate as other commodities. This is made clear by examining the Min and Max indices in Tables 1 and 2. Adding leisure as a ninth commodity will only change these indices if the ratio of leisure in period t to leisure in the base period is below the minimum or above the maximum for the other eight commodities. As a comparison of the Min columns reveals, leisure is often below the minimum for the other commodities (starred years in Table 2), and when it is not it is close to the minimum. This tends to reduce the growth in the full income indices. For example, Y_T is a weighted average of the ratios of current to base period consumption for each commodity. Adding leisure (consumption of which has been growing at a below average rate) reduces the average.

The economic factors underlying the differences in the growth rates of real income (without allowing for leisure) and real full income are as follows. For the eight purchased commodities, the income effect of rising real wages dominates any substitution effects due to changes in relative prices. Thus substantial increases in quantities consumed are observed over this period. However, for leisure the income effect of rising real wages is offset by the substitution effect due to the increase in the price of leisure relative to other commodities. Thus the growth in the consumption of leisure time is significantly lower than that of other commodities.

TABLE 1. Index Numbers of Real Income without Accounting for Leisure

Year	Q_L	Q_P	Q_F	Min x_t^i/x_0^i	Max x_t^i/x_0^i	Q_T
1949	100.0	100.0	100.0	100.0	100.0	100.0
1952	110.3	110.4	110.3	101.2	129.6	110.3
1956	130.7	131.2	131.0	115.0	175.4	127.2
1960	143.1	144.2	143.7	120.1	184.7	143.7
1964	151.9	152.0	151.9	127.5	216.9	152.0
1968	162.8	162.3	162.6	128.6	245.7	162.7
1972	182.5	183.4	182.9	108.4	294.6	183.1
1976	202.2	200.4	201.3	128.4	350.0	201.7
1980	194.6	193.3	194.0	126.8	350.0	194.2

TABLE 2. Index Numbers of the Standard of Living Accounting for Leisure

Year	Y_L	Y_P	Y_F	Y_T	Min x_t^i/x_0^i	Max x_t^i/x_0^i
1949	100.0	100.0	100.0	100.0	100.0	100.0
1952	105.8	105.6	105.7	105.7	101.2	129.6
1956	114.8	113.1	113.9	113.9	103.4*	175.4
1960	120.6	117.9	119.2	119.3	104.4*	184.7
1964	124.9	120.5	122.7	122.6	105.6*	216.9
1968	130.7	124.2	127.4	127.3	107.9*	245.7
1972	139.3	129.6	134.2	134.4	108.4	294.6
1976	150.1	134.6	142.1	141.6	112.9*	350.0
1980	147.5	134.5	140.9	140.4	113.9*	350.0

TABLE 3. Index Numbers of Real Wages and Real Non-Labour Income

Year	W_L	W_P	RWS	M_L	M_P
1949	100.0	100.0	100.0	100.0	100.0
1952	115.8	115.8	109.2	285.2	− 74.3
1956	140.5	144.0	125.6	568.7	− 63.1
1960	157.3	170.1	131.4	752.6	− 88.2
1964	170.7	184.2	143.4	890.1	− 60.3
1968	190.7	210.9	160.7	1,075.7	− 51.4
1972	217.6	265.5	186.4	1,349.0	− 52.6
1976	261.7	329.5	214.4	1,689.3	− 39.0
1980	256.6	341.5	205.4	1,608.7	− 51.0

TABLE 4. Index Numbers of Real Consumption Expenditure

Year	W_L	E_L	Q_L	Q_P
1949	100.0	100.0	100.0	100.0
1952	114.6	114.8	110.3	110.4
1956	137.6	137.2	130.7	131.2
1960	152.8	152.7	143.1	144.2
1964	164.9	163.9	151.9	152.0
1968	183.0	180.6	162.8	162.3
1972	207.5	205.4	182.5	183.4
1976	246.8	239.6	202.2	200.4
1980	241.9	233.0	194.6	193.3

Table 3 contains index numbers of real wages and real non-labour income. The expectation that M_P will often be negative and that the rate of increase of M_L will tend to be large is confirmed.

The W_L and W_P indices indicate that real wages have increased more than would be judged on the basis of traditional "real wage" measures that do not incorporate income-leisure choice. RWS is a comparable traditional real wage measure; it equals real wages and salaries and supplementary labour income per employee divided by the implicit price index for consumer expenditure.

Table 4 contains the Diewert consumption expenditure indices, and the comparable "real income" indices which do not incorporate leisure. As is clear from (69) and (70), the rate of growth of economic welfare is understated by the usual measures which do not account for the decline over time in hours worked. The magnitude of the understatement is indicated in Table 4. The growth in economic welfare was approximately 240 percent over the period 1949-76 rather than approximately 200 percent as the commonly used measures suggest.

5. Conclusions

This paper has accomplished two objectives; (1) to summarize and extend the economic theory of index numbers of the standard of living and the cost of living when allowance is made for joint consumption and labour supply decisions, and (2) to present some measures of real income and real wages which incorporate labour-leisure choice.

Most of the theoretical results in Section 3 on index numbers of economic welfare which incorporate income-leisure choice have analogous results for index numbers of the cost of living. Thus the the cost of living can be measured using full income, non-labour income, or the after-tax wage rate. The main reason for not including a section on the cost-of-living measures which incorporate leisure is that the concept of the cost of living seems limited to goods and services purchased in the marketplace whereas the concept of economic welfare clearly requires going beyond these and taking account of other factors affecting well-being.

While incorporating leisure time in measures of living standards is an important extension of existing measures, there are other extensions that would be worthwhile. One is to incorporate time spent in household production in addition to time allocated to leisure or working for pay. This third activity is particularly important to incorporate if the household is the basic unit of the analysis. A second is to allow non-labour income and wages to be endogenous; i.e., to incorporate savings decisions and decisions about education, training and other human capital investments.

Two *caveats* seem particularly important to mention. First, the theory of welfare measurement used in the paper applies to a single individual whereas the data employed are for the economy as a whole (albeit on a per capita basis). The welfare measures can be taken to apply to a single randomly selected member of the population. Second, the index number theory assumes that the individual is at a maximum utility position each period, or that the individual is working the desired number of hours. Both assumptions are commonly made, but this does not imply that they are unimportant. Measures which make a correction for the possibility that a randomly selected individual is working fewer weeks per year than desired could possibly be based on the aggregate unemployment rate.

Four different welfare measures which incorporate leisure were examined in the paper: full income, non-labour income, wage rate and consumption expenditure measures. The real wage and real consumption expenditure measures appear to be the most useful. Standard "real wage" and "real income" index numbers understate the increase in living standards; Tables 3 and 4 provide some indication of the degree of understatement.

Footnotes

[1] Throughout the paper I will refer to constant-utility index numbers as real income or cost-of-living index numbers and their observable counterparts as quantity or price indices.

[2] The term full income was introduced by Becker [1965].

References

Abbott, M. and O. Ashenfelter [1976]. "Labour Supply, Commodity Demand and the Allocation of Time", *Review of Economic Studies* 43, October 1976, pp.389-411.

Becker, G.S. [1965]. "A Theory of the Allocation of Time", *Economic Journal* 75, pp.493-517.

Braithwait, S.D. [1980]. "The Substitution Bias of the Laspeyres Price Index: An Analysis Using Estimated Cost-of-Living Indexes", *American Economic Review* 70, March 1980, pp.64-77.

Cleeton, D.L. [1982]. "The Theory of Real Wage Indices", *American Economic Review* 72, March 1982, pp.214-255.

Diewert, W.E. [1976]. "Exact and Superlative Index Numbers", *Journal of Econometrics* Vol. 4, May 1976, pp.115-145.

_____ [1981]. "The Economic Theory of Index Numbers: A Survey", in A. Deaton, ed., *Essays in the Theory and Measurement of Consumer Behavior in Honour of Sir Richard Stone*. Cambridge: Cambridge University Press, 1981.

_____ [1983]. "The Theory of the Cost-of-Living Index and the Measurement of Welfare Change", *Price Level Measurement: Proceedings From a Conference Sponsored by Statistics Canada.*

Lloyd, P.J. [1975]. "Substitution Effects and Biases in Nontrue Price Indices", *American Economic Review* 65, June 1975, pp.301-313.

_____ [1979]. "Constant-Utility Index Numbers of Real Wages: Comment", *American Economic Review* 69, September 1979, pp.682-685.

Noe, N.N. and G. von Furstenberg [1972]. "The Upward Bias in the Consumer Price Index Due to Substitution", *Journal of Political Economy* 80, November-December 1972, pp.1280-1294.

Pencavel, J.H. [1977]. "Constant Utility-Index Numbers of Real Wages", *American Economic Review* 67, March 1977, pp.91-100.

_____ [1979a]. "Constant-Utility Index Numbers of Real Wages: Revised Estimates", *American Economic Review* 69, March 1979, pp.240-243.

_____ [1979b]. "Constant-Utility Index Numbers of Real Wages: Reply", *American Economic Review* 69, September 79, pp.686-687.

Samuelson, P.A. and S. Swamy [1974]. "Invariant Economic Index Numbers and Canonical Duality: Survey and Synthesis", *American Economic Review*, 64, pp.566-593.

Usher, D. [1980]. *The Measurement of Economic Growth.* Oxford: Basil Blackwell, 1980.

PRICE LEVEL MEASUREMENT – W.E. Diewert (Editor)
Canadian Government Publishing Centre /
Elsevier Science Publishers B.V. (North-Holland)
© Minister of Supply and Services Canada, 1990

COMMENTS

John Bossons
Institute for Policy Analysis
University of Toronto

The questions raised by Jack Triplett's paper for this conference apply to the paper by Craig Riddell and I think it is useful to consider Riddell's paper in the light of Triplett's key question: What is the purpose of an index? Before I do so, however, I would like to make one very general point. Much has been said at this conference about the relative advantages of different indexes and the possible benefits associated with different ways of correcting indexes. Particularly when thinking in terms of the applicability of cost of living indexes to tax correction and labour bargaining and in the development of indexed financial contracts, the familiarity of the existing Consumer Price Index is an important advantage. Because changes may reduce the credibility of an accepted index, there may be important costs associated with changing a widely used index. These costs have to be offset against any potential benefits to be derived from such change. I want to stress this point. Looking at it from the viewpoint of using such indexes in tax correction and the development of indexed financial instruments, the stability of the definition of the CPI and its general acceptance are very important.

I have a number of comments on Craig Riddell's paper. A basic point I want to make is that the really interesting question that Riddell has raised is whether there is any gain to be obtained by developing a new "current period" cost-of-living index which is more "complete" than the Consumer Price Index. The question is a "second best" problem, and can be rephrased as follows:

Given that a fully-complete cost-of-living index is ideal but not implemented, is it better to use an incomplete index (a "subindex" in the terms used in Diewert's paper to this conference) rather than another? As any student of the second-best problem will appreciate, a "more complete" subindex is not necessarily "better" in the sense of providing a closer approximation to the unimplemented ideal.

It is important to emphasize first that we are talking about one subindex versus another. No current-period index can be a complete cost-of-living index without taking account of intertemporal allocation. Indeed, contrary to what Riddell has claimed, it is not true that the intertemporal problem arises only if savings is introduced. Even ignoring savings, there are important intertemporal allocation problems inevitably involved in labour-leisure choices. Consequently, I will claim that it is wrong to think that one can easily go from what we can now easily measure, namely a current-period subindex for consumption goods, to anything that is very much further on the road to completeness. I will argue that those difficulties are very relevant in any attempt to define a "complete" current-period subindex.

There are several problems that arise in dealing with labor-leisure choices. First, a major complication is introduced by the necessity of dealing with involuntary leisure (i.e. involuntary unemployment). This complication is empirically important. One has to be very careful not to invent implicit welfare measures which "show" that welfare was historically maximized in periods such as 1933. An increase in unemployment is an increase in welfare only if one assumes (as in the Lucas-Rapping model) that all unemployment is voluntary. But if one assumes that the temporary Keynesian-type disequilibria can occur in which people get stuck with involuntarily-reduced income and an involuntarily-chosen level of unemployment, a serious problem arises. The introduction of involuntary unemployment necessarily introduces intertemporal planning problems. Involuntary unemployment can be viewed as a constraint on the intertemporal planning problem which reduces welfare. There is a corresponding shadow price which can be viewed as a measure of the value of the availability of work given uncertainty regarding the availability of work. This uncertainty arises both with respect to whether existing job will be continued in subsequent periods or, for somebody who is not now working and is currently consuming leisure, whether leisure taken in the current period can be switched at the consumer's discretion into time spent at work during subsequent periods. The intertemporal labor allocation problem cannot be appropriately characterized without taking risk into account.

Ignoring the risk element introduced by unemployment leaves out a very important component of the problem of constructing a full welfare measure for a current period. This component is correspondingly a component of the problem of defining a "complete" current-period cost-of-living index that reflects the consumer's full cost-minimization problem. I want to stress this point in order to emphasize that it is a non-trivial problem to specify how the currently measured goods price subindex can be expanded into a more complete cost-of-living index. Erwin Diewert has emphasized (in Section 9 of his paper) the difficulties in going beyond a complete current-period subindex to define index numbers that correspond to the intertemporal allocation problem. I'd submit that the difficulties of going from the current CPI index or prices of consumption goods and servies to a "full" current-period subindex are equally as great.

Raising the problems created by involuntary unemployment naturally leads to the associated problem of whose leisure time to take as weights in trying to measure a full cost-of-living index. Should a cost-of-living index vary as the age composition of the population changes, changing the average amount of leisure taken through retirement by the population as a whole? Should a cost-of-living index vary with the aggregate level of unemployment, changing as measured labour force participation ratios change in response to the discouraged worker effects? Contrary to the position taken by Riddell, I don't believe that it is straightforward to develop measures of the cost-of-living that incorporate changes in the price of leisure.

Further, I want to argue that incorporating adjustments for changes in labour-leisure choices will for many applications not even be appropriate. Here I want to repeat some points that Jack Triplett made in his paper. For example, in dealing with tax indexation issues, it is necessary to ask what is the purpose of trying to define an index for the purpose of indexing tax rates, exemptions, and so forth. I submit that to a considerable extent what is needed for this purpose is a measure of the trade-off between private and public consumption goods and services. One way of arguing this is to think of this in terms of a utility tree which implies that decisions about the consumption of public goods (including redistributive transfers) are made for given choices by the population about how time expected to be available over their lives will on average be allocated between leisure and time spent in work to create the goods and services which are going to be divided up between private consumption goods and public consumption goods. If one thinks of it that way, then it's not inappropriate to use for such purposes a current-period subindex for consumption goods and services.

I would argue that the errors introduced because of the difficulties that arise in trying to construct a complete single current-period measure decrease the signal-to-noise ratio for the Index relative to what one now has with the currently-used subindex. Even if it were preferable (ignoring those measurement problems) to use a more complete current-period subindex, for what purposes is Riddell's characterization of the problem relevant?

His key assumptions are that income is endogenous and leisure choice is voluntary. I think Riddell's analysis is very useful for the longer-run questions which he addressed towards the end of his paper, such as the welfare measurement problem earlier introduced by Tobin that arises in correcting the national accounts to obtain a measure that more appropriately measures long-run changes in welfare. However, this analysis of secular trends in welfare is a very different kind of issue from the application that is implied by using an index for purposes of tax corrections, wage escalation, and other short run measurement problems.

Now I would like to make a couple of comments on some broader questions which are partly stimulated by Jack Triplett's paper as well as by Riddell's paper. I want to argue that a very important application of the consumer price index is in its potential role in correcting the tax system for inflation and in the development of indexed financial instruments. Here the problem is to obtain a "correct" measure of real income from capital, which can alternatively be defined as obtaining a measure of the real trade-off between future and present consumption of goods and services.

I think the simplest way to characterize the relevant consumer optimization problem in this application is in a life cycle context in which the most important price is the price index for consumption goods during retirement, at which point in the life cycle a person can be assumed to be 100 percent engaged in leisure. With this simplification, the relevant consumer intertemporal allocation problem can be characterized as one in which the leisure/labor choice is not relevant. While obviously excluding an important aspect of the complete intertemporal problem, I believe this characterization is relevant for the purpose of getting at the critical indexing problem that applies when one is defining a "correct" definition of taxable income from capital, the appropriate definition of income from capital when one is constructing an indexed financial instrument, of the appropriate index to be used in indexing a life annuity or pension. I would consequently argue that a current-period subindex for consumption goods is very close to the correct index that should be used for this purpose, so that the consumer price index or some variant is, in fact, "the right index" for this purpose.

The example provided by the application to defining a "correct" measure of income from capital for purposes of taxation or for defining financial contracts and pensions in real terms illustrates the importance of the points emphasized by Triplett. The appropriate characterization of the relevant consumer allocation problem differs depending on the application, and this has important implications for the choice of index. Depending on the application, a current-period subindex for consumption goods and services may be a better measure than a "complete" cost-of-living index even ignoring the measurement problems that would arise if the "complete" index were to be used. Where this is the case, the "second best" problem analyzed in Riddell's paper becomes irrelevant.

PRICE LEVEL MEASUREMENT – W.E. Diewert (Editor)
Canadian Government Publishing Centre /
Elsevier Science Publishers B.V. (North-Holland)
© Minister of Supply and Services Canada, 1990

PREFERENCE DIVERSITY AND AGGREGATE ECONOMIC COST-OF-LIVING INDEXES[1]

Charles Blackorby and David Donaldson
Department of Economics
University of British Columbia

SUMMARY

For a single household, an economic cost-of-living index compares the minimum expenditure needed to achieve a particular indifference curve under two different price regimes.

We are interested in a price index for a group of households rather than a single one. Further, it is likely that preferences, even if homothetic, are different for different households. It would be useful to be able to combine the individual price indexes into an aggregate price index that is itself capable of being given a reasonable economic interpretation. Two general possibilities exist, the non-optimizing and optimizing approaches.

The non-optimizing approach attempts to find conditions under which aggregate demand behaviour could be generated by a "representative" household. Preferences of this representative household are then used to construct an expenditure function, which is, in turn, used to construct a cost-of-living index. The alternative optimizing procedure is to construct a social expenditure function based on a (Bergson-Samuelson) social-welfare function and individual preferences. This social expenditure function is then used to construct a social cost-of-living index. In this paper, we pursue both of these procedures, and attempt to find conditions under which the resulting aggregate cost-of-living index is reference-level free.

1. Introduction

For a single household, an economic cost-of-living index compares the minimum expenditure needed to achieve a particular indifference curve under two different price regimes.

More specifically, let the preferences of a particular household (h) be represented by the expenditure function e^{h2}, so that e^h (u_h,p) is the minimum expenditure needed by the household to achieve the level of welfare indexed by u_h at prices p = $(p_1,...,p_m)$. A (Konüs-type) cost-of-living index for household h is defined to be

$$I^h (p^1, p^0, \bar{u}_h) = \frac{e^h (\bar{u}_h, p^1)}{e^h (\bar{u}_h, p^0)} \; , \tag{1}$$

the ratio of minimum expenditures at "comparison" prices p^1 to minimum expenditures at "reference" prices p^0. This index is homogeneous of degree one in comparison prices and homogeneous of degree minus one in reference prices. The reference level of well-being is indexed by $\bar{u}_h$. It may refer to a given indifference surface of the household, or it may be the level of utility generated by a particular commodity bundle $\bar{x}^h$, with

$$\bar{u}_h = U^h (\bar{x}^h) \tag{2}$$

where U^h is a (direct) utility function corresponding to e^h.[3]

For many purposes, however, we want the index I^h to be independent of the reference level of utility, $\bar{u}_h$. For this to be true, it is necessary and sufficient that the preferences of household h be homothetic. That is, U^h can be written as

$$U^h (x^h) = \overset{*}{U}{}^h (\bar{U}^h (x^h)), \tag{3}$$

where $\overset{*}{U}{}^h$ is increasing and $\bar{U}^h$ is positively linearly homogeneous. Equivalently, the preferences are homothetic if and only if the expenditure function e^h can be written[4] as

$$e^h (u_h, p) = \Pi^h (p) \phi^h (u_h) \tag{4}$$

where Π^h is positively linearly homogeneous and ϕ^h is increasing. In this case the index I^h is reference-level free,[5] and

$$I^h (p^1, p^0, \bar{u}_h) = \frac{\Pi^h (p^1)}{\Pi^h (p^0)}. \tag{5}$$

In general, we are interested in a price index for a group of households rather than a single one. Further, it is likely that preferences, even if homothetic, are different for different households. It would be useful to be able to combine the individual price indexes into an aggregate price index that is itself capable of being given a reasonable economic interpretation. Two general possibilities exist, the non-optimizing and optimizing approaches.

The non-optimizing approach attempts to find conditions under which aggregate demand behaviour could be generated by a "representative" household (Gorman [1953, 1961]). Preferences of this representative household are then used to construct an expenditure function, which is, in turn, used to construct a cost-of-living index. The alternative optimizing procedure (Pollak [1980, 1981]) is to construct a social expenditure function based on a (Bergson-Samuelson) social-welfare function and individual preferences. This social expenditure function is then used to construct a social cost-of-living index.

In this paper, we pursue both of these procedures, and attempt to find conditions under which the resulting aggregate cost-of-living index is reference-level free. In Section 2, we discuss the existence of a representative consumer and the corresponding index. First we look at the case of exogenously given prices and arbitrary distributions of purchasing power. The resulting index is independent of the reference level of utility of the representative household if and only if the economy is composed of households which have identical homothetic preferences. We then investigate the same question with a potentially less demanding price domain. We allow first some, and then all prices to be determined endogenously in a general equilibrium setting. Unfortunately, this relaxation produces no new solutions, leaving us again with identical homothetic preferences. We also investigate briefly (following Lau [1977], Jorgenson, Lau and Stoker [1981, 1982]) a situation in which household expenditure functions depend on attributes other than utility levels. Again, this relaxation leads to no new generalizations; all households must have identical homothetic preferences if the aggregate cost-of-living index is to be reference-level free.

In Section 3, we turn to the maximizing approach. Following Pollak [1980, 1981], we construct a social expenditure function. It is the minimum aggregate expenditure, allowing lump-sum redistributions of purchasing power, necessary to achieve a given level of social welfare. We then use it to construct a **social** cost-of-living index and ask that the index be reference-level free (independent of the chosen level of social welfare). We note that identical homothetic preferences produce a reference-level-free social cost-of-living index, but we are principally concerned with the possibility that different households (or population subgroups) may have different preferences. We consider arbitrary linear social-welfare functions (thus restricting our attention to additively separable social-welfare functions). Our main result is that it is possible to find a social cost-of-living index that is reference-level free when households have homothetic but non-identical preferences. This aggregate index depends, in general, on the weights in the social-welfare function. These indexes are easily computed from individual unit expenditure functions, ($\{\Pi^h(p)\}$) and we provide formulas. We discuss the relationship of our indexes to the Prais [1959] – Muellbauer [1974] – Nicholson [1975] – Pollak [1980] share-weighted average and equally-weighted average of individual price indexes.

It is also possible to ask if the social-welfare function could be chosen to rationalize the existing distribution of purchasing power, and we show how it can be done. We further discuss, in the concluding remarks, the possibility of choosing weights to find the social weights implicit in the choice of a particular cost-of-living index.

2. Representative Households

In this section, we explore the problem of the existence of a representative household under different price regimes. In the first case, prices are exogenously given, in the second, some or all of the prices are determined endogenously, and in the third, households are identified by special characteristics such as age of head, region and so on.

2.1 Exogenous Prices

Here, we review the well-known case explored by Gorman [1953]. We present it in some detail to make further exposition clear.

Each household faces prices p with income $y_h > 0$, and we may write aggregate demand for good j as

$$x_j (p, y_1,...,y_H) = \sum_h x_j^h(p, y_h),$$
(6)

$j = 1,...,m$. This places no restrictions on the aggregate demand functions. However, we are interested in finding conditions under which the aggregate demands of (6) could be generated by a single rational agent whose income is $Y = \sum_h y_h$. Thus, we require

$$x_j (p, y_1,...,y_H) = \bar{x}_j (p, Y) = \sum_h x_j^h (p, y_h),$$
(7)

for a demand function $\bar{x}_j$, $j = 1,...,m$. Since the representative household is assumed to be rational, it must have an expenditure function. If its indifference curves are indexed by w, then we must have, since $Y = \sum_h y_h$,

$$e (w, p) = \sum_h e^h (u_h, p)$$
(8)

for all p. By setting $p = (1,...,1) = 1_m$ we get

$$e (w, 1_m) = \sum_h e^h (u_h, 1_m),$$
(9)

and with a simple normalization of w,

$$w = \sum_h \psi^h (u_h).$$
(10)

Since equation (8) is an identity, it may be differentiated, using Shephard's lemma, to obtain

$$\hat{x}_j (w, p) = \sum_h \hat{x}_j^h (u_h, p)$$
(11)

for all p, where $\hat{x}_j$ and $\{\hat{x}_j^h\}$ are the compensated demand functions for good j for the representative and given households. Substitution of indirect utility functions[6] into (11) yields (7), thus reducing the problem to finding the conditions under which

$$e \left(\underset{h}{\Sigma} \psi^h (u_h), p \right) = \underset{h}{\Sigma} e^h (u_h, p) \tag{12}$$

can hold for all p and for all $u = (u_1, \ldots, u_H)$.

We want to know what restriction (12) places on the individual and representative expenditure functions. The solution to this problem is well known and due originally to Gorman [1953]. There exists a representative household if and only if each household's expenditure function may be written as

$$e^h (u_h, p) = \Pi (p) \psi^h (u_h) + \beta^h (p), \tag{13}$$

and

$$e (w, p) = \Pi (p) w + \beta (p), \tag{14}$$

where w is given by (10) and

$$\beta (p) := \underset{h}{\Sigma} \beta^h (p). \tag{15}$$

This means, since Π is independent of h in (13), that the Engel curves for the households must be parallel straight lines. The aggregate cost-of-living index

$$I (p^1, p^0, \bar{w}) = \frac{e (\bar{w}, p^1)}{e (\bar{w}, p^0)} = \frac{\Pi (p^1) \bar{w} + \beta (p^1)}{\Pi (p^0) \bar{w} + \beta (p^0)}, \tag{16}$$

and, in order for it to be reference-level free, we must have

$$\beta (p) = \sum_h \beta^h (p) = c \, \Pi(p), \tag{17}$$

for some real number c. In this case,

$$I (p^1, p^0, \bar{w}) = \frac{\Pi (p^1)}{\Pi (p^0)}. \tag{18}$$

The representative household's preferences are homothetic, but (strictly speaking) each household does not necessarily have homothetic preferences. (14) allows "committed" expenditures by households, and (17) requires that non-proportional committed expenditures sum to zero. Thus some committed expenditures must be positive and others **negative** (not allowing (14) to hold everywhere) if (17) holds without all committed expenditures zero. For this reason, and the fact that these differences in taste are very sensitive to changes in the number of households, we choose to ignore the differences in taste associated with these expenditures and set them all equal to zero. In this case each household has identical homothetic preferences, and its reference-level-free cost-of-living index is the appropriate one for the group.

This result is very restrictive. It results from the requirement that the distribution of income (and hence of utilities) be arbitrary, at least in some open region of income space, and the requirement that all prices be exogenous.

2.2 Endogenously Determined Prices

We therefore attempt to generalize Gorman's result by allowing some or all of the prices to be determined endogenously by the interaction of supply and demand. To do this, we partition the list of commodities $J = (1,...,m)$ into two groups; $\hat{J}$ is the set of commodities whose prices are exogenously given, and $\overset{o}{J}$ is the set whose prices are endogenously determined. Thus

$$J = (\hat{J}, \overset{o}{J}) \tag{19}$$

and we write the corresponding prices as

$$p = (\hat{p}, q). \tag{20}$$

Since, by Shephard's lemma, the price derivatives of expenditure functions are compensated demand functions, we may write our requirement as

$$e_j (w, \hat{p}, q) = \sum_h e_j^h (u_h, \hat{p}, q), \; j \, \epsilon \, \hat{J} \tag{21}$$

and

$$e_j (w, \hat{p}, q) = \sum_h e_j^h (u_h, \hat{p}, q) = S^j (\hat{p}, q), \; j \, \epsilon \, \overset{o}{J}, \tag{22}$$

where e_j and e_j^h are the derivatives of e and e^h with respect to the jth price. S^j is the supply function for the jth good. (21) and (22) must hold only for price-utility combinations that are compatible with the price determination in (22). We assume that (22) can be solved to get

$$q_j = Q^j (\hat{p}, w), \; j \, \epsilon \, \overset{o}{J}, \tag{23}$$

and we assume that the solution is unique (in the case when $\hat{J}$ is empty, we select a *numeraire*). We could, alternatively, solve this system of equations as

$$w = \Gamma^j (\hat{p}, q_j), \; j \, \epsilon \, \overset{o}{J}. \tag{24}$$

Setting $\hat{p} = (1,...,1)$ in (23) allows $\hat{p}$ and q to be substituted out of (21) and (22). We assume that these equations can be solved uniquely for w as

$$w = W(u) = W(u_1,..., u_H). \tag{25}$$

W must be increasing in its arguments in order that aggregate expenditure respond positively to individual expenditures.

Equations (21) and (22) imply, using Euler's theorem and the positive linear homogeneity of expenditure functions in prices, that

$$e\,(w,\,\hat{p},\,q) = \sum_h e^h\,(u_h,\,\hat{p},\,q), \qquad (26)$$

or that (using (23) and (25))

$$e(W(u),\,\hat{p},\,Q\,(\hat{p},\,W\,(u))) = \sum_h e^h(u_h,\,\hat{p},\,Q(\hat{p},W(u))). \qquad (27)$$

for all $\hat{p}$ and u, where

$$Q(\hat{p},\,w) = \left(\{Q^j(\hat{p},\,w)\}_{j\,\epsilon\,\overset{o}{J}} \right). \qquad (28)$$

Equation (27) is similar to equation (12), but there are several differences. First, we cannot be sure that W is additive in household utilities. Second, the domain of (26) is restricted by the endogenous pricing. In fact, (21) and (22) are **not** implied by (26) unless there are no endogenous prices. We require (21) and (22) to be satisfied, and, therefore, satisfaction of (26) is necessary but not sufficient. It would seem intuitively clear, however, that these new conditions would allow more scope for individual variability than before. That this is **not** the case is stated in

Theorem 1: Given our regularity conditions, (21) and (22) hold for all $\hat{p}$, u if and only if the expenditure functions can be written as

$$e^h(u_h,\,p) = \Pi\,(p)\,\psi^h\,(u_h) + \beta^h\,(p), \qquad (29)$$

and

$$e(w,\,p) = \Pi(p)\,w + \beta(p), \qquad (30)$$

where

$$\beta(p) := \sum_h \beta^h (p) \tag{31}$$

and

$$w = W(u) = \sum_h \psi^h (u_h). \tag{32}$$

Proof: See the appendix.

In the proof, we show that W must satisfy a particular functional restriction, namely that (with a harmless normalization),

$$w = \sum_h F^h (u_h, w), \tag{33}$$

where F^h is increasing in u_h. Subsequently, we show that (21) and (22) require that F^h not depend on w. Thus, no change in preferences from the solution to Gorman's problem has been found. Hence, there exists a reference-level-free aggregate cost-of-living index if and only if all households have identical homothetic preferences. The cost-of-living index is given by (18).

2.3 Households with Multiple Characteristics

In recent years, several articles[7] have investigated aggregate demand equations which are identified by various attributes or characteristics. This allows a representative household to exist with preferences depending on some aggregate values of those characteristics. This has led to representative preferences that are more general than the Gorman restriction on preferences (above) and are easy to implement.

We pursue this notion briefly. Suppose that each household has a level of a characteristic indexed by v_h. Then we require

$$e(w_1, w_2, p) = \sum_h e^h (u_h, v_h, p) \tag{34}$$

for all p, $\{u_h\}$ and $\{v_h\}$. Setting $p = 1_m$ in (34) requires that the aggregate characteristics w_1 and w_2 depend on u and v. We require that

$$w_1 = W^1 (u,v) \tag{35}$$

and

$$w_2 = W^2 (u, v). \tag{36}$$

The solution to this problem is known (Gorman [1978]) to be given by

$$e^h(u_h, v_h, p) = {}^1\Pi (p)\psi^{h1} (u_h, v_h) + {}^2\Pi(p)\psi^{h2} (u_h, v_h) + \beta^h (p), \tag{37}$$

and

$$e(w_1, w_2, p) = {}^1\Pi (p) w_1 + {}^2\Pi (p) w_2 + \beta (p), \tag{38}$$

where

$$w_1 = W^1(u,v) = \sum_h \psi^{h1} (u_h, v_h), \tag{39}$$

$$w_2 = W^2(u,v) = \sum_h \psi^{h2} (u_h, v_h), \tag{40}$$

and

$$\beta(p) := \sum_h \beta^h(p). \tag{41}$$

The aggregate cost-of-living index in this case is

$$\bar{I}(p^1, p^0, w_1, w_2) = \frac{{}^1\Pi(p^1)\, w_1 + {}^2\Pi(p^1)\, w_2 + \beta(p^1)}{{}^1\Pi(p^0)\, w_1 + {}^2\Pi(p^0)\, w_2 + \beta(p^0)}, \qquad (42)$$

and we want it to be independent of w_1 and w_2. The conditions for this are

$$^1\Pi(p) = c_1\, \Pi(p), \qquad (43)$$

$$c_1 > 0,$$

$$^2\Pi(p) = c_2\, \Pi(p), \qquad (44)$$

$$c_2 \geq 0,$$

$$\beta(p) = c_3\, \Pi(p), \qquad (45)$$

for all p. Thus

$$e(w_1, w_2, p) = \Pi(p)\,(c_1\, w_1 + c_2\, w_2 + c_3). \qquad (46)$$

In each of these cases, households must have identical homothetic preferences. In constructing aggregate cost-of-living indexes, nothing has been gained.

3. The Social Cost-of-Living Index

Pollak [1980, 1981] has proposed a procedure for finding an aggregate cost-of-living index based on a social-welfare function. It is derived from a social expenditure function – the minimum expenditure required (with lump-sum redistributions) to achieve a given level of social welfare.

Following the usual practice, we assume that social welfare is

$$w = W(u) \tag{47}$$

and that W is continuous, differentiable, and quasi-concave. We now need the assumption that each expenditure function is **strictly convex** in u_h. This rules out degenerate solutions, and is equivalent to the assumption that each household's direct utility function has a strictly concave representation.[8]

We define the social expenditure function as

$$\bar{e}(w, p) = \inf_u \left\{ \sum_h e^h(u_h, p) \mid W(u) \geq w \right\}. \tag{48}$$

We want to limit our attention to additively separable social-welfare functions, where

$$w = \sum_h a_h \, \psi^h(u_h), \tag{49}$$

and each ψ^h is concave. A simple normalization[9] of individual utilities makes this into

$$w = \sum_h a_h u_h. \tag{50}$$

We assume that no a_h is negative and that at least one is positive. In this case, the social expenditure function is

$$e(w, p, a) = \inf_u \left\{ \sum_h e^h(u_h, p) \mid \sum_h a_h u_h \geq w \right\}, \tag{51}$$

and it depends on the level of social welfare, w, the price vector p, and the vector of social weights a.

The social cost-of-living index is defined in the usual manner, using the social expenditure function e. The ratio of expenditures necessary to achieve social-welfare level w in the two price regimes is

$$I(p^1, p^0, w, a) = \frac{e(w, p^1, a)}{e(w, p^0, a)}. \tag{52}$$

We would like to discover the conditions under which this cost-of-living index is independent of w. It is easy to show that, for this to be true it is necessary and sufficient that the social expenditure function can be written as

$$e(w, p, a) = \Lambda(p,a) \, T(w, a), \tag{53}$$

implying that

$$I(p^1, p^0, w, a) = \frac{\Lambda(p^1, a)}{\Lambda(p^0, a)}. \tag{54}$$

The social expenditure function must be decomposable into two functions. One depends on prices and social weights a, the other depends on w and the social weights. In general, the social cost-of-living index, even if reference-level free, depends on a, and in the special case that it does not, individual preferences will again prove to be identical and homothetic.

The domain of u_h in e^h is an interval of the real line (or, perhaps, all of it), and we denote this as D^h. Again we rule out precommitted expenditures (as in Section 2) and require

$$\inf\{e^h(u_h, p) \mid u_h \, \epsilon \, D^h\} = 0 \,.^{10} \tag{55}$$

By setting $a_g = 0$ for all $g \neq h$, $a_h = 1$, we note that (55) requires

$$e(w,p,(0,...,1,...,0)) = e^h(w, p). \tag{56}$$

Thus, satisfaction of (53) for all a requires

$$e^h(u_h,p) = \Pi^h(p) \, \psi^h(u_h). \tag{57}$$

Therefore, a necessary condition (given (55)) for reference-level freedom of the social cost-of-living index is that households have (possibly non-identical) homothetic preferences. Further, if a single household is the only one that counts in social welfare, its cost-of-living index is the social cost-of-living index.

To proceed with the analysis of possibilities for individual preferences, we prove a simple social duality result. We denote the optimal value of u_h in (51) as u_h^*, so that, whenever these values exist,

$$
\begin{aligned}
e(w,p,a) &= \min\left\{ \sum_h e^h(u_h, p) \mid \sum_h a_h u_h \geq w \right\} \\
&= \sum_h e^h(u_h^*, p).
\end{aligned}
\tag{58}
$$

Lemma 1: If $u_h^* \ \varepsilon$ int D^h for all h, then

$$
u_h^* = - \frac{e_{a_h}(w,p,a)}{e_w(w,p,a)}
\tag{59}
$$

where the subscripts denote partial derivatives.

Proof: See the appendix.

We may use the special structure on $e(w,p,a)$ in (53) to place structure on the optimal level of utilities.

$$
\begin{aligned}
u_h^* &= - \frac{\Lambda_{a_h}(p,a)\, T(w,a) + \Lambda(p,a)\, T_{a_h}(w,a)}{\Lambda(p,a)\, T_w(w,a)} \\
&= - \left[\frac{\Lambda_{a_h}(p,a)}{\Lambda(p,a)}\right]\left[\frac{T(w,a)}{T_w(w,a)}\right] - \left[\frac{T_{a_h}(w,a)}{T_w(w,a)}\right] \\
&:= \alpha^h(p,a)\,\Gamma(w,a) + \Phi^h(w,a)
\end{aligned}
\tag{60}
$$

Our main results are derived from (60) and the first-order conditions for problem (51).

Several different possibilities arise, depending on the nature of the functions in (53) and (60). If the function is independent of a, the cost-of-living index is independent of the social weights. This suggests immediately that individual preferences must be identical as well as homothetic. Thus, we are taken back to the representative household of Section 2.

There are other solutions, however. They allow different homothetic preferences for the households. They correspond to the cases where $\Gamma(w, a)$ is independent of w, and where $\Phi^h(w, a)$ is independent of w. Our main results are contained in

Theorem 2: Given our regularity conditions,[11] the social cost-of-living index is reference-level free for all a > 0, if and only if individual households' expenditure functions can be written as one of

Case 1:

$$e^h(u_h, p) = \Pi(p)\psi^h(u_h),\tag{61}$$

where ψ^h is a strictly convex function, $h = 1,...H,$

Case 2:

$$e^h(u_h, p) = \Pi^h(p) \exp\{u_h\}\tag{62}$$

where D^h is $\mathbb{R}$, $h = 1,...,H,$

or

Case 3:

$$e^h(u_h, p) = \Pi^h(p)\left[r(u_h - c_h)\right]^{\frac{1}{r}},\tag{63}$$

where $D^h = \{u_h \mid u_h \geq c_h\}, 0 < r < 1,$

and $D^h = \{u_h \mid u_h \leq c_h\}, r < 0, h = 1,...,H.$

Corresponding to these household preferences, the social cost-of-living indexes are

Case 1:

$$I(p^1, p^0, w, a) = \frac{\Pi(p^1)}{\Pi(p^0)},\tag{64}$$

Case 2:

$$I(p^1, p^0, w, a) = X_h \left[\frac{\Pi^h(p^1)}{\Pi^h(p^0)}\right]^{\bar{a}_h},\tag{65}$$

where $\bar{a}_h := a_h / \sum_g a_g,$ and

Case 3:

$$I(p^1, p^0, w, a) = \frac{\left[\sum_h a_h^{\frac{1}{1-r}} \Pi^h(p^1)^{\frac{-r}{1-r}}\right]^{\frac{1-r}{-r}}}{\left[\sum_h a_h^{\frac{1}{1-r}} \Pi^h(p^0)^{\frac{-r}{1-r}}\right]^{\frac{1-r}{-r}}}\tag{66}$$

Proof: See the appendix.

These results merit some discussion. Case 1 is identical homothetic preferences, and the social cost-of-living index is the common individual index. Of course, the social cost-of-living index is the common individual one for any social-welfare function whatever. Cases 2 and 3 allow non-identical preferences. In Case 2, the social cost-of-living index is a Cobb-Douglas function of the individual price indexes. If everyone receives an equal weight, then the social cost-of-living index is the geometric mean of the individual indexes. This contrasts with the "democratic" price index of Prais and Nicholson (generalized by Muellbauer and Pollak). It is an arithmetic mean of individual price indexes. The index in Case 3 is a ratio of weighted means of order $(-r/(1-r))$ of the individual "unit" cost functions $\Pi^h(p)$. $(-r/(1-r))$ can take on all real values between $-\infty$ and one excluding zero). The limiting case of zero is just Case 2. As r approaches $-\infty$, the index becomes

$$\frac{\sum\limits_h \Pi^h(p^1)}{\sum\limits_h \Pi^h(p^0)} = \frac{\frac{1}{H}\sum\limits_h \Pi^h(p^1)}{\frac{1}{H}\sum\limits_h \Pi^h(p^0)} , \qquad (67)$$

the ratio of means of unit costs. This is closely related to, but not equal to the democratic price index.

The above indexes may be rewritten in terms of optimal expenditure shares, $\left\{s^*_h\right\}$. For household h, in Case 2,

$$s^*_h = \frac{a_h}{\sum\limits_g a_g} = \bar{a}_h . \qquad (68)$$

Thus, the social cost-of-living index can be rewritten as

$$I(p^1, p^0, w, a) = \underset{h}{X}\left(\frac{\Pi^h(p^1)}{\Pi^h(p^0)}\right)^{s^*_h} \qquad (69)$$

In Case 3, the optimal expenditure share for household h depends on the price vector p as well as the set of weights, and

$$s^*_h = S^h(p, a) = \frac{a_h^{\frac{1}{1-r}} \; \Pi^h (p)^{-\frac{r}{1-r}}}{\sum_g a_g^{\frac{1}{1-r}} \; \Pi^g (p)^{-\frac{r}{1-r}}} , \qquad (70)$$

and the social cost-of-living index can be shown to be

$$I (p^1, p^0, w, a) = \frac{\left[\sum_h S^h (p^1, a)^{(1-r)} \; \Pi^h (p^1)^r \right]^{\frac{1}{r}}}{\left[\sum_h S^h (p^0, a)^{(1-r)} \; \Pi^h (p^0)^r \right]^{\frac{1}{r}}} . \qquad (71)$$

(70) and (71) suggest that the indexes we have proposed are related to the "plutocratic" index of Prais and Nicholson (the base-period-share-weighted average of individual Laspeyres indexes) and its generalizations by Pollak (the same average of arbitrary household indexes).[12] However, the mean that appears here is not an ordinary arithmetic mean, and the unit cost functions rather than the individual indexes are used in the averages.

Practical use of these indexes demands that an important question be discussed. That is the question of choosing a and r (Case 2 corresponds to r = 0). r is an "inequality-aversion" parameter. This may be most easily seen by computing the Bergson-Samuelson indirect utility functions for Cases 2 and 3 (we set $c_h = 0$ in Case 3 for simplicity). In each case, we simply find $\sum a_h u_h$ with the indirect utility functions substituted in. Since in Case 2,

$$e^h(u_h, p) = \Pi^h(p) \exp\{u_h\} = y_h \leftrightarrow u_h = \log \frac{y_h}{\Pi^h (p)} , \qquad (72)$$

we have found the individual indirects, and $\sum a_h u_h$ is equivalent to

$$B(p, y_1,...,y_H) = \sum_h a_h \log \left(\frac{y_h}{\Pi^h(p)} \right) \tag{73}$$

$$= \log \left[\underset{h}{X} \left(\frac{y_h}{\Pi^h(p)} \right)^{a_h} \right].$$

In Case 3, a similar computation yields

$$B(p, y_1,...,y_H) = \sum_h \frac{a_h}{r} \left(\frac{y_h}{\Pi^h(p)} \right)^r, r < 1, r \neq 0. \tag{74}$$

Thus, the Bergson-Samuelson indirects are the quasi-concave CES-Cobb- Douglas family of functions, the (weighted) means of order r (r < 1), and the independent variables are individual real incomes. r = 1 represents no inequality aversion, r = - ∞ is "maximin" – complete inequality aversion. The weights a_h can be set normatively, and in this case they would presumably be equal (differences in need ought to appear in the unit cost functions $\{\Pi^h(p)\}$. In this case, the social cost-of-living indexes can be computed directly from (65) and (66), presumably for several values of r.

On the other hand, it is possible (for any r) to choose the a's to rationalize any pattern of expenditure shares. Thus it would be possible to use actual expenditure shares to justify a choice of a's. This presumes implicitly that the government is maximizing a social-welfare function with (possibly peculiar) weights. If this approach is taken, the social cost-of-living indexes can be computed directly from (69) and (71).

This procedure suggests another exercise. Suppose that we have estimated a unit cost function for each income class in the economy, treating each group as if it consisted of identical households. We can then try different values of a and r to see which ones best fit the actual cost-of-living indexes constructed by government agencies. In this way, another "inverse-optimal" problem could be solved.

4. Conclusion

We have investigated two general procedures for finding reference-level-free aggregate cost-of-living indexes. The non-maximizing approach involves the search for a representative consumer, and we have shown that this approach results in an unhelpful solution. All households must have identical homothetic preferences, with the aggregate index equal to the individual indexes.

The maximizing framework allows more variability in individual preferences since the distribution of purchasing power is adjusted to suit the dictates of a social-welfare function in finding the social cost-of-living index. We have investigated the case of additively separable social-welfare functions only, and have found all the sets of individual preferences that will produce reference-level-free social cost-of-living indexes for arbitrary social weights. Individual households must have homothetic preferences, but they may differ from household to household. The social cost-of-living indexes are ratios of weighted means of individual unit cost functions. These weighted means are all in the CES-Cobb-Douglas family and contain, in addition to the social weights, an inequality-aversion parameter.

Appendix

Theorem 1: Given our regularity conditions, (21) and (22) hold for all $\hat{p}$, u if and only if the expenditure functions can be written as

$$e^h(u_h, p) = \Pi(p)\, \psi^h(u_h) + \beta^h(p), \tag{29}$$

and

$$e(w, p) = \Pi(p)\, w + \beta(p), \tag{30}$$

where

$$\beta(p) := \sum_h \beta^h(p) \tag{31}$$

and

$$w = W(u) = \sum_h \psi^h (u_h).$$ (32)

Proof: We consider two cases, first the case where $|\overset{o}{J}| = 1$, and second the case where $|\overset{o}{J}| = m$.

Case 1: Equation (27) may be differentiated with respect to u_h to yield

$$e_w W_h + e_q \frac{\partial q}{\partial u_h} = e_u^h + \sum_g e_q^g \frac{\partial q}{\partial u_h}.$$ (a.1)

Using (22), this becomes

$$e_w(W(u), \hat{p}, Q (W(u), \hat{p})) W_h(u) = e_u^h (u_h, \hat{p}, Q (W(u), \hat{p})).$$ (a.2)

Since the right-hand side of this equation depends on $\hat{p}$, u_h and w only, so does the left. Consequently, we can write

$$W_h (u) = f^h (u_h, W(u)),$$ (a.3)

and

$$e_u^h (u_h, \hat{p}, Q(w, \hat{p})) = e_w(w, \hat{p}, Q(w, \hat{p})) f^h(u_h, w).$$ (a.4)

Substituting $q = Q(w, \hat{p})$ and $w = \Gamma(\hat{p}, q)$ (24) in (a.4),

$$e_u^h (u_h, \hat{p}, q) = e_w (\Gamma(\hat{p}, q), \hat{p}, q) f^h (u_h, \Gamma(\hat{p}, q))$$ (a.5)

$$=: \alpha (\hat{p}, q) f^h(u_h, \Gamma(\hat{p}, q)).$$

Integrating,

$$e^h(u_h, \hat{p}, q) = \alpha(\hat{p}, q) \, F^h \, (u_h, \, \Gamma(\hat{p}, q)) + \beta^h(\hat{p}, q) \tag{a.6}$$

and

$$e \, (w, \, \hat{p}, q) = \alpha(\hat{p}, q) \sum_h \left[F^h \, (u_h, \, \Gamma(\hat{p}, q)) + \beta^h(\hat{p}, q) \right] \tag{a.7}$$

$$= \alpha(\hat{p}, q) \, F \, (w) + \beta(p, q).$$

F(w) may be replaced by w in (a.7) by using a harmless normalization. Testing (a.6) and (a.7) against (21) and (22) yields $F^h_\Gamma = 0$, and hence (29) and (30).

Case 2: In this case, (a.4) becomes

$$e^h_u \, (u_h, \, \bar{Q} \, (w)) = e_w \, (w, \, \bar{Q} \, (w)) \, f^h \, (u_h, \, w) \tag{a.8}$$

where $\bar{Q}$ is the vector of endogenously determined prices. Since $w = \Gamma^j \, (q)$ for each j, (a.7) becomes

$$e^h(u_h, \, q) = \alpha(q) \, F^h \, (u_h, \, \Gamma^1(q), \, \Gamma^2(q), ..., \Gamma^m(q)) \tag{a.9}$$

$$+ \, \beta^h(q).$$

An analogous argument to the one above makes

$$F^h(u_h, \, \Gamma^1(q), ..., \, \Gamma^m(q)) = \, \psi^h \, (u_h). \ \square$$

Lemma 1: If $u^*_h \ \varepsilon \ int \ D^h$ for all h, then

$$u_h^* = - \frac{e_{a_h}(w,p,a)}{e_w(w,p,a)} \tag{59}$$

Proof: Since

$$e(w,p,a) = \sum_h e^h(u_h^*, p) \tag{a.10}$$

for all w and a,

$$e_{a_h}(w,p,a) = \sum_h e_u^h(u_h^*,p) \frac{\partial u_h^*}{\partial a_h} . \tag{a.11}$$

Since $u_h^* \ \varepsilon \ \text{int } D^h$,

$$e_u^h(u_h^*,p) = \lambda a_h \tag{a.12}$$

where λ is a Lagrange multiplier. Hence,

$$e_{a_h}(w,p,a) = \lambda \sum_h a_h \frac{\partial u_h^*}{\partial a_h} . \tag{a.13}$$

Because

$$\sum_h a_h u_h^* = w \tag{a.14}$$

for all a,

$$\sum a_h \frac{\partial u_h^*}{\partial a_h} + u_h^* = 0. \tag{a.15}$$

Consequently

$$e_{a_h} (w,p,a) = -\lambda u_h^* . \qquad (a.16)$$

By a similar argument,

$$e_w(w,p,a) = \lambda , \qquad (a.17)$$

and since $\lambda \neq 0$, the conclusion follows. $\square$

Lemma 2: If $u_h^* \; \varepsilon$ int D^h and $a_h > 0$ for all h,

$$\frac{\partial u_h^*}{\partial w} > 0.$$

Proof: The first-order conditions for u_h^* are

$$e_u^h (u_h^*, p) = \lambda a_h \qquad (a.19)$$

where λ is a Lagrange multiplier. This holds for all w, and so

$$e_{uu}^h (u_h^*, p) \frac{\partial u_h^*}{\partial w} = \frac{\partial \lambda}{\partial w} a_h , \qquad (a.20)$$

$h = 1,...,H$. Since e is strictly convex in u, $e_{uu}^h (u_h^*, p) > 0$. Hence the sign of $\partial u_h^*/\partial w$ must be the same for all w. Since $a_h > 0$, they must all be positive. $\square$

Theorem 2: Given our regularity conditions, the social cost-of-living index is reference-level free for all $a > 0$ if and only if individual households' expenditure functions may be written as

Case 1:

$$e^h(u_h, p) = \Pi(p)\psi^h(u_h), \tag{61}$$

where ψ^h is a strictly convex function in $h = 1,...,H$,

Case 2:

$$e^h(u_h, p) = \Pi^h(p) \exp\{u_h\} \tag{62}$$

where D^h is $\mathbb{R}$, $h = 1,...,H$,

Case 3:

$$e^h(u_h, p) = \Pi^h(p)\left[r(u_h - c_h)\right]^{\frac{1}{r}}, \tag{63}$$

where $D^h = \{u_h \mid u_h \geq c_h\}$, $0 < r < 1$,

and $D^h = \{u_h \mid u_h \leq c_h\}$, $r < 0$, $h = 1,...,H$.

Corresponding to these household preferences, the social cost-of-living indexes are

Case 1:

$$I(p^1, p^0, w, a) = \frac{\Pi(p^1)}{\Pi(p^0)}, \tag{64}$$

Case 2:

$$I(p^1, p^0, w, a) = \underset{h}{X} \left[\frac{\Pi^h (p^1)}{\Pi^h (p^0)} \right]^{\bar{a}_h}, \tag{65}$$

where $\bar{a}_h: = a_h / \underset{g}{\Sigma} a_g$,

Case 3:

$$I(p^1, p^0, w, a) = \frac{\left[\underset{h}{\Sigma} a_h^{\frac{1}{1-r}} \Pi^h(p^1)^{\frac{-r}{1-r}} \right]^{\frac{1-r}{-r}}}{\left[\underset{h}{\Sigma} a_h^{\frac{1}{1-r}} \Pi^h(p^0)^{\frac{-r}{1-r}} \right]^{\frac{1-r}{-r}}} \tag{66}$$

Proof: Using our regularity condition, we choose $a = \bar{a} >> 0$ such that u_h^* ε int D. Thus,

$$u_h^* = \bar{\alpha}^h (p, \bar{a}) \, \Gamma(w, \bar{a}) + \psi^h (w, \bar{a}). \tag{60}$$

We may suppress $\bar{a}$ in (60) by writing it as

$$u_h^* = \bar{\alpha}^h (p) \, \bar{\Gamma} (w) + \psi^h (w). \tag{a.21}$$

Since u_h^* ε int D^h and since

$$e^h(u_h, p) = \Pi^h (p) \, \psi^h (u_h). \tag{57}$$

$$\Pi^h (p) \, \psi^{h'} (u_h^*) = \lambda \, \bar{a}_h. \tag{a.22}$$

From the proof of Lemma 1,

$$e_w(w,p,a) = \lambda , \qquad\qquad (a.17)$$

and from (53),

$$e_w(w,p,\bar{a}) = \Lambda(p, \bar{a})\, T_w\,(w,\bar{a}) \qquad\qquad (a.23)$$

$$=: \bar{\Lambda}(p)\, k\,(w).$$

Using (a.17) and (a.23), (a.22) may be rewritten as

$$\frac{\psi^{h\,\prime}\,(u_h^*)}{\bar{a}_h} = \frac{\bar{\Lambda}\,(p)\,k(w)}{\Pi^h(p)} . \qquad\qquad (a.24)$$

Defining $f^h\,(t) := \psi^{h\,\prime}\,(t)\,\bar{a}_h$, (a.24) can be rewritten, using (a.21)

$$f^h(\bar{\alpha}^h(p)\,\bar{\Gamma}\,(w) + \psi^h(w)) = \frac{\bar{\Lambda}(p)}{\Pi^h(p)}\,k(w). \qquad\qquad (a.25)$$

By Lemma 2, we know that u_h^* is increasing in w, since e_{uu}^h is positive, f^h is increasing. Therefore, k is increasing in w.

(a.25) admits of several solutions.

Case 1: Suppose that $\bar{a}^h\,(p)$ is constant. Then the left-hand side is independent of p, and the right-hand side must be as well. Thus, $\Pi^h\,(p) = c_h\bar{\Lambda}\,(p)$ with

$$e^h(u_h, p) = c_h\,\bar{\Lambda}\,(p)\,\psi^h\,(u_h) \qquad\qquad (a.26)$$

which is equivalent to (61).

(61) implies that

$$e(w,p,a) = \inf\left\{\sum_h \Pi(p) \, \psi^h(u_h) \mid \sum_h a_h u_h \geq w\right\}$$

$$= \Pi(p) \inf\left\{\sum_h \psi^h(u_h) \mid \sum_h a_h u_h \geq w\right\} \qquad (a.27)$$

$$= \Pi(p) \, T(w,a).$$

Now suppose that $\bar{a}^h(p)$ is not a constant and denote it as x. Then

$$\frac{\bar{\Lambda}(p)}{\Pi^h(p)} = g^h(x), \qquad (a.28)$$

so that (a.25) becomes

$$f^h(x \, \bar{\Gamma}(w) + \bar{\psi}^h(w)) = g^h(x) \, k(w). \qquad (a.29)$$

Case 3: Now suppose $\bar{\psi}^h(w) = c_h$, a constant.

Then

$$f^h(x \, \bar{\Gamma}(w) + c_h) = g^h(x) \, k(w). \qquad (a.30)$$

$\Gamma(w)$ must be increasing in w in this case (since u^*_h is), and we denote it as z. Then (a.30) becomes

$$\bar{f}^h(xz) = g^h(x) \, \bar{k}(z), \qquad (a.31)$$

a Pexider equation. x is positive, but z is not restricted in sign. The solution is (Eichhorn [1978], Blackorby and Donaldson [1982])

$$\bar{f}^h(t) = \begin{cases} \bar{b}_h \mid t \mid^\sigma, t \ge 0, \bar{b}_h > 0, \\ -\hat{b}_h \mid t \mid^\sigma, t \le 0, \hat{b}_h > 0. \end{cases} \tag{a.32}$$

Consequently,

$$\psi^{h\prime}(t) = \begin{cases} \bar{d}_h \mid t - c_h \mid^\sigma, t \ge c_h, \bar{d}_h > 0, \\ -\hat{d}_h \mid t - c_h \mid^\sigma, t \le \hat{c}_h, \hat{d}_h > 0. \end{cases} \tag{a.33}$$

and

$$\psi^h(u_h) = \begin{cases} \dfrac{\bar{d}_h}{\sigma+i} \mid u_h - c_h \mid^{\sigma+1}, u_h > c_h, \bar{d}_h > 0, \\ \dfrac{-d_h}{\sigma+1} \mid u_h - c_h \mid^{\sigma+1}, u_h < c_h, \hat{d}_h > 0. \end{cases} \tag{a.34}$$

Increasingness requires $\sigma + 1 \ne 0$, and strict convexity requires $(\sigma + 1) \ge 1$ for $u_h \ge c_h$, and $(\sigma + 1) \le 1$ for $u_h \ge c_h$. However, we must have $\psi^h_{uu}(u_h) > 0$, so that $0 < (\sigma + 1) < 1$ is ruled out. Consequently, ψ^h can be written as

$$\psi^h(u_h) = d_h \left(r(u_h - c_h) \right)^{\frac{1}{r}}, \tag{a.35}$$

$0 < r < 1$ and $D^h = \{u_h \mid u_h \ge c_h\}$,

or $r < 0$ and $D^h = \{u_h \mid u_h \le c_h\}$.

In both cases, $d_h > 0$. This yields (63), Case 3, and a simple computation gives (66).

Case 2: Now suppose that $\Gamma(w) = c$, a constant. In this case, (a.29) becomes

$$f^h(cx + \bar{\phi}^h(w)) = g^h(x) k(w). \tag{a.36}$$

Defining $w: = cx$ and $z: = \bar{\phi}^h_h (w)$ (ϕ is increasing since u^*_h is increasing), we get

$$f^h (y + z) = \hat{g}^h (y) \hat{k} (z). \qquad (a.37)$$

This is a Pexider equation whose solution is (Eichhorn [1978])

$$f^h(t) = c_h e^t, \qquad (a.38)$$

so that

$$\psi^{h'} (u_h) = \frac{c_h e^{u_h}}{a_h} \qquad (a.39)$$

and

$$\psi^h (u_h) = b_h e^{u_h}. \qquad (a.40)$$

This yields (62)(Case 2) and (65) follows from a simple computation.

We must now show that there are no other possibilities. If Γ and ϕ^h are both sensitive to w, then (a.29) holds, and we get, by differentiating,

$$f^{h'} (x \bar{\Gamma}(w) + \phi^h (w)) \bar{\Gamma}(w) = g^{h'} (x) k (w) \qquad (a.41)$$

and

$$f^{h'} (x \bar{\Gamma} (w) + \phi^h (w) (x \bar{\Gamma}'(w) + \phi^{h'} (w)) = g^h (x) k'(w). \qquad (a.42)$$

The left-hand side of (a.39) is not zero for all w, so $g'(x) \neq 0$ for some w. Further, we know $k (w) > 0$. Therefore,

$$x\; \frac{\bar{\Gamma}'(w)}{\bar{\Gamma}(w)} + \frac{\bar{\phi}h'(w)}{\bar{\Gamma}(w)} = \frac{g^h(x)}{g^{h'}(x)}\; \frac{k'(w)}{k(w)} \tag{a.43}$$

or

$$x\hat{\Gamma}(w) + \hat{\gamma}^h(w) = \hat{g}^h(x)\hat{k}(w). \tag{a.44}$$

If $\hat{\Gamma}(w)$ is constant, we get an immediate impossibility. If not, let $\hat{\Gamma}(w) = z$, so that (a.42) becomes

$$x\,z + \bar{\gamma}^h(z) = g^h(x)\,\bar{k}(z). \tag{a.43}$$

Differentiating,

$$z = \hat{g}^{h'}(x)\,\bar{k}(z) \tag{a.44}$$

requiring $g^{h'}(x) = $ a constant, $\hat{g}^h(x) = c\,x$ and $\bar{k}(z) = z/c$. Thus, (a.43) requires $\gamma^h(z) = 0$, implying $\bar{\phi}^{h'}(w) = 0$, a contradiction. $\square$

Footnotes

[1] We are indebted to Bert Balk, Erwin Diewert, and Margaret Slade for helpful comments and criticisms.

[2] We assume that prices are all positive, and that e^h satisfies certain regularity properties: e^h is, (i) non-negative, (ii) jointly continuous in (u_h, p), (iii) concave, positively linearly homogenous, and non-decreasing in p, (iv) continuously differentiable in (u_h, p), (v) twice continuously differentiable in u_h, (vi) increasing in u_h, and (vii) strictly convex in u_h. (vii) is not needed until Section 3.

[3] e^h and U^h must represent the same preferences. Thus, $e^h(u_h, p) = \min_{x^h}\{\Sigma p_j x_j^h \mid U^h(x^h) \geq u_h\}$. See Diewert [1974].

[4] See Blackorby, Primont and Russell [1978, Lemma 3.4].

[5] See Pollak [1982].

[6] These are easily found from expenditure functions since e $(u, p) = y \leftrightarrow V (p, y) = u$.

[7] See Lau [1977], Jorgenson, Lau and Stoker [1981, 1982].

[8] See Diewert [1978] for a discussion.

[9] Make $\bar{u}_h = \psi^h(u_h)$ so that $\bar{e}_h(\bar{u}_h, p): = e^h(\psi^{h^{-1}}(\bar{u}_h), p)$. Since ψ^h is concave, $\psi^{h^{-1}}$ is convex, and since e^h is strictly convex in u_h, $\bar{e}^h$ is strictly convex in $\bar{u}_h$. The reasons for additive separability are mathematical tractability and the fact that we usually want indexes that apply to any number of people.

[10] Without this assumption, we would get precommitted expenditures of $\beta^h(p)$ for person h, and reference-level freedom would require $\Sigma_h \beta^h(p) = 0$ for all p.

[11] We use a harmless additional regularity condition, namely that for all w in the interior of D (its domain in e), there exists $\bar{a} \gg 0$ such that $u_h^* \epsilon$ int D^h for all h.

[12] called the Scitovsky-Laspeyres index. See Pollak [1980] for a discussion.

References

Blackorby, C. and D. Donaldson, "Ratio-Scale and Translation-Scale Full Interpersonal Comparability Without Domain Restrictions: Admissible Social Evaluation Functions", *International Economic Review*, 23, 1982, pp.249-268.

_____ , D. Primont and R. Russell, *Duality, Separability and Functional Structure: Theory and Economic Applications*, Elsevier/North-Holland, 1978.

Diewert, W.E., "Applications of Duality Theory", in *Frontiers of Quantitative Economics*, II, (M. Intriligator and D. Kendrick, eds.) North-Holland, 1974.

_____ , "Hicks' Aggregation Theorem and the Existence of a Real Value-Added Function", in *Production Economics: A Dual Approach to Theory and Applications*, Volume 2, (M. Fuss and D. McFadden, eds.), North-Holland, 1978.

Eichhorn, W., *Functional Equations in Economics*, Addison-Wesley, 1978.

Gorman, W., "Community Preference Fields", *Econometrica* 21, 1953, pp.63-80.

_____ , "On a Class of Preference Fields", *Metroeconomica*, 13, 1961, pp.53-56.

_____ , "More Measures for Fixed Factors", London School of Economics Mimeo, 1978.

Jorgenson, D., L. Lau and T. Stoker, "Aggregate Consumer Behavior and Individual Welfare", in *Macroeconomic Analysis* (D. Currie, R. Nobay and D. Peel, eds.), Croom-Helm, 1981.

_____ , "The Transcendental Logarithmic Model of Aggregate Consumer Behavior", in *Advances in Econometrics* V.1 (R. Baseman and G. Rhodes, eds.) JAI, 1982.

Lau, L., "Existence of Conditions for Aggregate Demand Functions: The Case of Multiple Indexes", Technical Report 249, IMSSS, Stanford 1977.

Muellbauer, J., "The Political Economy of Price Indices", Discussion Paper 22, Birkbeck College, 1974.

Nicholson, J., "Whose Cost-of-Living?", *Journal of the Royal Statistical Society*, Part 4, 138, 1975, pp.540-542.

Pollak, R., "Group Cost-of-Living Indexes", *American Economic Review*, 70, 1980, pp.273-278.

————— , "The Social Cost-of-Living Index", *Journal of Public Economics*, 15, 1981, pp.311-336.

————— , "The Theory of the Cost-of-Living Index", *Price Level Measurement: Proceedings From a Conference Sponsored by Statistics Canada*, 1983.

Prais, S., "Whose Cost-of-Living?", *Review of Economic Studies*, 26, 1959, pp.126-134.

PRICE LEVEL MEASUREMENT – W.E. Diewert (Editor)
Canadian Government Publishing Centre /
Elsevier Science Publishers B.V. (North-Holland)
© Minister of Supply and Services Canada, 1990

AXIOMATIC FOUNDATION OF PRICE INDEXES AND PURCHASING POWER PARITIES

Wolfgang Eichhorn and
Joachim Voeller
Institut für Wirtschaftstheorie und Operations Research
Universistät Karlsruhe*

SUMMARY

The paper is concerned with the so-called "axiomatic" ("statistical", "mechanistic") approach to the construction of price indexes or purchasing power parities. In this branch of index theory the theoretical foundations of both intertemporal and interspatial price comparisons are built on certain basic assumptions, called axioms, which are meant to be so general as to be satisfied by all relevant "mechanistic" price indexes/purchasing power parities. Whereas, historically, much controversy existed in the literature about the usefulness, consistency, and independence of certain sets of requirements for, generally speaking, index functions, today many propositions or theorems shed light on the various interrelations between a great number of different sets of properties for index functions. Also several characterizations of certain well-known price indexes/purchasing power parities have been published and offer an excellent chance to judge the quality of the characterized function. In this context characterization means the deduction of a certain index function from a set of given conditions such that the function in question not only meets the required conditions but also represents the only function that satisfies the conditions.

Section 1 of the paper gives a short introduction into the "axiomatic approach". Section 2 then offers an axiomatic definition of the concept of a mechanistic price index/purchasing power parity. Also, the independence of the definitional set of axioms is proven. A few well-known examples of price indexes/purchasing power parities as well as basic implications of the definitional set of axioms are presented in Section 3. Further criteria and implications are introduced in Section 4 which are partially used in Section 5 to characterize Fisher's "Ideal Index" and the so-called "Cobb-Douglas Index". It is obvious that conflicting requirements lead to inconsistent sets of conditions. Problems relating to such inconsistencies are analysed in the final Section 6.

1. Introduction

The relationship between the purchasing power of money and the level of prices can be studied either over time or space or even with respect to both dimensions. In the case of purely **intertemporal** price comparisons, the purchasing power of money in a certain geographical area, usually a country, is defined to be inversely related to the value of some appropriately chosen **price index**. Thus, price comparisons over time essentially deal with the methodological and practical problems concerning the construction and the application of price and/or quantity indexes. **Interspatial** price comparisons, however, analyse the relationship between the price level and the value of money in two different places at a certain point of time. Though methodologically quite similar to time-to-time comparisons, place-to-place investigations pose additional difficulties both in theory and practice. In all interspatial comparisons the concept of **purchasing power parity** is used to measure the price level in one location relative to that in another. Generally, the prices are collected for those goods and services that represent most adequately particular consumption or production patterns in two countries. Therefore a purchasing power parity between two countries indicates the number of currency units of one country equivalent in purchasing power to one currency unit of the other country.

It is obvious that purchasing power parities can also be computed between two different places in a certain country, for instance, between two different cities. In this case the purchasing power parity compares the purchasing power of the same currency at two locations. Also, the fact should be noted that in certain methods for multi-country comparisons the respective prices of all countries involved influence the value of a purchasing power parity between any two countries.

The following sections set out what is commonly called the "axiomatic" ("statistical", "mechanistic") approach to the construction of price indexes or purchasing power parities. The theoretical foundations of this branch of index theory are built on certain basic assumptions, called axioms, which are meant to be so general as to be satisfied by all relevant price indexes/purchasing power parities in national or international comparisons over time or space. Historically, much controversy existed in the literature (see, for instance, Eichhorn/Voeller [1976], Voeller [1982]) about the usefulness, consistency and independence of certain sets of requirements for, generally speaking, index functions. Today, however, numerous propositions or theorems shed light on the various interrelations between a great number of different sets of properties for index functions. Also several characterizations of certain well-known price indexes/purchasing power parities have been published and offer an excellent chance to judge the quality of the characterized function. In this context characterization means the deduction of a certain index function from a set of given conditions such that the function in question not only meets the required conditions but also represents the only function that can be deduced from these conditions.

In closing this brief introduction we take a closer look at the contents of Sections 2-7. In Section 2 an axiomatic definition of the concept of a price index/purchasing power parity is given and the independence of the definitional set of axioms is proven. Besides a few well-known examples of price indexes/purchasing power parities Section 3 presents basic implications. Further criteria and implications are introduced in Section 4 which are partially used in Section 5 to characterize two famous indexes. It is obvious that conflicting requirements will lead to inconsistent sets of conditions, and problems relating to such inconsistencies are investigated in Section 6. Throughout the paper many abbreviated references are made to the literature which can be found in detail in the Bibliography in Section 7.

2. Notation and Definitions

It is assumed that n goods and services of identical or equivalent quality are priced either

- at two different points of time in the same place or
- in two different locations (usually two different countries) at the same point of time.

The assumption of a common list of commodities is made for simplicity reasons. Clearly, the identification of equivalent items requires

– that the time interval between the two points of time is not too long;
– that the two countries do not fall too far apart as far as their levels of economic and social development are concerned,

respectively. In both cases the selection of the commodities from the respective production or consumption pattern must not impair their representativeness, that is, the goods and services chosen should adequately represent either the situation at two different points of time or in two distinct locations (countries).

Let

$$q^A = (q_1^A,...,q_n^A) \in \mathbb{R}_{++}^n \quad \left(\mathbb{R}_{++} := \left\{r \mid r \in \mathbb{R}, r > 0\right\}\right)$$

$$q^B = (q_1^B,...,q_n^B) \in \mathbb{R}_{++}^n$$

be the vectors of the quantities or weights of the n items purchased

– at a **base time A** and a **comparison time B** or
– at a **base country (location) A** and a **comparison country (location) B**,

respectively. The corresponding prices of these commodities are given by the price vectors

$$p^A = (p_1^A,...,p_n^A) \in \mathbb{R}_{++}^n$$

$$p^B = (p_1^B,...,p_n^B) \in \mathbb{R}_{++}^n .$$

Note that in the case of comparisons between countries the prices are given in different currencies. We point out here that throughout our paper the prices and quantities are **independent** variables. In this case one speaks of **mechanistic** or **statistical** price indexes. If the preferences of the households are taken into consideration, prices and quantities **depend on each other**. Then one speaks of **economic** price indexes or **cost-of-living** indexes. The theory of these indexes is independently developed in this volume by Diewert and Pollak. Both of them deduce connections between mechanistic and economic price indexes.

(2.0) **Definition**:

A function

$$P: \mathbb{R}^{4n}_{++} \to \mathbb{R}_{++}, \ (q^A, p^A, q^B, p^B) \mapsto P(q^A, p^A, q^B, p^B)$$

is called a **mechanistic (statistical) price index** (in the case of intertemporal comparisons) or a **purchasing power parity for country B with respect to country A** (in the case of interspatial comparisons) if P satisfies the following four axioms (2.1) to (2.4) for all $(q^A, p^A, q^B, p^B) \in \mathbb{R}^{4n}_{++}$. Then the value $P(q^A, p^A, q^B, p^B)$ represents the value of the price index or the purchasing power parity at the price-quantity situation (q^A, p^A, q^B, p^B).

Systems of axioms for cost-of-living indexes can be found in Diewert's, and Pollak's, contributions to this volume.

(2.1) **Monotonicity Axiom**:

The function P is strictly increasing with respect to p^B and strictly decreasing with respect to p^A:

$$P(q^A, p^A, q^B, p^B) > P(q^A, p^A, q^B, \bar{p}^B) \quad \text{if } p^B \geq \bar{p}^{B,'}$$

$$P(q^A, p^A, q^B, p^B) < P(q^A, \bar{p}^A, q^B, p^B) \quad \text{if } p^A \geq \bar{p}^A.$$

(2.2) **Proportionality Axiom**:

If all corresponding prices differ by the same factor λ ($\lambda \in \mathbb{R}_{++}$), then the value of the function P equals λ:

$$P(q^A, p^A, q^B, \lambda p^A) = \lambda \quad (\lambda \in \mathbb{R}_{++}).$$

(2.3) Price Dimensionality Axiom:

The same proportional change in the unit of the currency (currencies) does not change the value of the function P:

$$P(q^A, \lambda p^A, q^B, \lambda p^B) = P(q^A, p^A, q^B, p^B) \qquad (\lambda \in \mathbb{R}_{++}).$$

(2.4) Commensurability Axiom:

The same change in the units of measurement of the corresponding commodities does not change the value of the function P:

$$P(\frac{q_1^A}{\lambda_1}, \dots, \frac{q_n^A}{\lambda_n}, \lambda_1 p_1^A, \dots, \lambda_n p_n^A, \frac{q_1^B}{\lambda_1}, \dots, \frac{q_n^B}{\lambda_n}, \lambda_1 p_1^B, \dots, \lambda_n p_n^B)$$

$$= P(q^A, p^A, q^B, p^B) \qquad (\lambda_1, \dots, \lambda_n \in \mathbb{R}_{++}).$$

Conditions (2.1) to (2.4) are called axioms since all four requirements are economically reasonable and thus represent basic properties which are desirable for every price index or purchasing power parity. Compliance with this definitional set of axioms then implies that the function P in question sensitively registers either the change of a country's price level or the purchasing power of one country's currency with respect to the other country's money.

At this point attention is called to the fact that a certain price index/purchasing power parity P is based on a given basket of goods. If, for instance, $n = 1$, then P: $\mathbb{R}_{++}^4 \rightarrow \mathbb{R}_{++}$ satisfying (2.1) to (2.4) is given by the so-called **price relative** or **price ratio** of the particular commodity, i.e.,

$$P(q_1^A, p_1^A, q_1^B, p_1^B) = \frac{p_1^B}{p_1^A}. \qquad (2.5)$$

The proof is trivial since the Proportionality Axiom (2.2) gives

$$P(q_1^A, p_1^A, q_1^B, p_1^B) = P(q_1^A, p_1^A, q_1^B, \frac{p_1^B}{p_1^A} p_1^A) = \frac{p_1^B}{p_1^A}$$

which satisfies (2.1) to (2.4). $\square$

Hence, in the special case of only one commodity, Axiom (2.2) is sufficient to characterize the index function. It is obvious that the shape of a function P cannot be determined so easily in the case of $n \geq 2$. In fact, the set or class of all functions satisfying (2.1) to (2.4) is not known yet. Some examples will be presented in the next section but before that, it is shown that the set of Conditions (2.1) to (2.4) is not redundant.

(2.6) Theorem:

Let $n \geq 2$. Axioms (2.1) to (2.4) are independent in the following sense: Any three of these axioms can be satisfied by a function P which does not satisfy the remaining axiom.

Proof:

The function denoted by

$$P(q^A, p^A, q^B, p^B) = \left(\frac{p_1^B}{p_1^A}\right)^{\alpha_1} \left(\frac{p_2^B}{p_2^A}\right)^{\alpha_2} \cdots \left(\frac{p_n^B}{p_n^A}\right)^{\alpha_n} \tag{2.7}$$

$$(\alpha_1 < 0, \alpha_2 > 0, \dots \alpha_n > 0 \text{ real constants}, \Sigma \alpha_i = 1)$$

fulfills Axioms (2.2), (2.3), (2.4), but not Axiom (2.1). The function given by

$$P(q^A, p^A, q^B, p^B) = \sum_{i=1}^{n} \frac{p_i^B}{p_i^A} \tag{2.8}$$

satisfies Axioms (2.1), (2.3), (2.4), but not Axiom (2.2). The function represented by[2]

$$P(q^A, p^A, q^B, p^B) = \frac{q^A p^A}{q^A p^A + 1} \frac{1}{n} \sum_{i=1}^{n} \frac{p_i^B}{p_i^A} + \frac{1}{q^A p^A + 1} \max \left\{ \frac{p_1^B}{p_1^A}, \ldots, \frac{p_n^B}{p_n^A} \right\} \qquad (2.9)$$

meets Axioms (2.1), (2.2), (2.4), but not Axiom (2.3). (Formula (2.9) was developed by H. Funke.) Finally the function given by

$$P(q^A, p^A, q^B, p^B) = \frac{\sum p_i^B}{\sum p_i^A} \qquad (2.10)$$

conforms to Axioms (2.1), (2.2), (2.3), but not to Axiom (2.4). $\square$

Evidently, the functions (2.7), (2.8), (2.9) and (2.10) cannot be regarded as price indexes/purchasing power parities in the sense of Definition (2.0).

3. Examples and Implications

As examples of price indexes/purchasing power parities, i.e., functions satisfying Axioms (2.1) to (2.4) the following well-known indexes[3] are presented:

$$P(q^A, p^A, q^B, p^B) = \frac{q^A p^B}{q^A p^A} \qquad \text{(``Laspeyres index''); } \qquad (3.1)$$

$$P(q^A, p^A, q^B, p^B) = \frac{q^B p^B}{q^B p^A} \qquad \text{(``Paasche index ''). } \qquad (3.2)$$

The Laspeyres index using weights of time/country A unfortunately neglects time/country B's weights (consumption or production pattern). On the other hand, the Paasche index using weights in B does not consider the weights in A. As a compromise the geometric mean of the two indexes was favored by Fisher [1927]:

$$P(q^A, p^A, q^B, p^B) = \left[\frac{q^A p^B}{q^A p^A} \cdot \frac{q^B p^B}{q^B p^A} \right]^{\frac{1}{2}} \qquad \text{(``Fischer's ideal index'')}; \qquad (3.3)$$

of which

$$P(q^A, p^A, q^B, p^B) = \left(\frac{q^A p^B}{q^A p^A} \right)^{\alpha} \cdot \left(\frac{q^B p^B}{q^B p^A} \right)^{1-\alpha} \qquad (0 \le \alpha \le 1) \qquad (3.4)$$

is a generalization. The convex combination

$$P(q^A, p^A, q^B, p^B) = \alpha \frac{q^A p^B}{q^A p^A} + (1-\alpha) \frac{q^B p^B}{q^B p^A} \qquad \begin{array}{l} (0 \le \alpha \le 1; \text{``Drobisch} \\ \text{index'' for } \alpha = \frac{1}{2}) \end{array} \qquad (3.5)$$

of the indexes (3.1) and (3.2) generalizes the arithmetic mean of the Laspeyres and Paasche index.

Average weights, as applied in most of the following examples, indicate a compromise situation, too.

$$P(q^A, p^A, q^B, p^B) = \frac{(\alpha q^A + (1-\alpha) q^B) p^B}{(\alpha q^A + (1-\alpha) q^B) p^A} \qquad \begin{array}{l} (0 \le \alpha \le 1; \\ \text{``Marshall-Edgeworth} \\ \text{index'' for } \alpha = \frac{1}{2}). \end{array} \qquad (3.6)$$

Note that the functions given by (3.4) to (3.6) represent the Laspeyres index (3.1) for $\alpha = 1$ and the Paasche index for $\alpha = 0$.

$$P(q^A,p^A,q^B,p^B) = \frac{\sum\limits_{i=1}^{n} \sqrt{q_i^A q_i^B}\, p_i^B}{\sum\limits_{i=1}^{n} \sqrt{q_i^A q_i^B}\, p_i^A} \qquad \text{(``Walsh index'')}; \qquad (3.7)$$

$$P(q^A,p^A,q^B,p^B) = \frac{\sum\limits_{i=1}^{n} \dfrac{q_i^A\, q_i^B}{q_i^A + q_i^B}\, p_i^B}{\sum\limits_{i=1}^{n} \dfrac{q_i^A\, q_i^B}{q_i^A + q_i^B}\, p_i^A} \qquad \begin{array}{l}\text{(``Geary-Khamis-}\\ \text{index'')};\end{array} \qquad (3.8)$$

$$P(q^A,p^A,q^B,p^B) = \left(\frac{p_1^B}{p_1^A}\right)^{\alpha_1} \left(\frac{p_2^B}{p_2^A}\right)^{\alpha_2} \cdots \left(\frac{p_n^B}{p_n^A}\right)^{\alpha_n} \qquad \begin{array}{l}\text{(``Cobb-Douglas}\\ \text{index'')}\end{array} \qquad (3.9)$$

$$(\alpha_1 > 0,\ldots,\alpha_n > 0 \text{ real constants}, \Sigma\alpha_i = 1).$$

For the assessment of Definition (2.0) the following theorem offers further arguments. It shows that the fulfillment of Axioms (2.1) to (2.4) leads to compliance with additional conditions which all make good economic sense. In accordance with ``Irving Fisher's test approach'' these conditions are called tests. Fisher [1927] used his tests to derive price index formulas and to test their quality.

(3.10) Theorem:

Every price index/purchasing power parity, i.e., every function P satisfying Axioms (2.1) to (2.4) also meets the following tests (3.11) to (3.14).

(3.11) Identity Test:

If all corresponding prices do not differ in absolute terms (but possibly in denomination), then the value of the function P equals 1:

$$P(q^A,p^A,q^B,p^A) = 1.$$

(3.12) Weak Proportionality Test:

If the quantity vectors satisfy $q^A = q^B$ and if all corresponding prices differ by the same factor λ ($\lambda \in \mathbb{R}_{++}$), then the value of the function P equals λ:

$$P(q^A, p^A, q^A, \lambda p^A) = \lambda. \qquad (\lambda \in \mathbb{R}_{++}).$$

(3.13) Quantity Dimensionality Test:

The same proportional change in all quantity units (weights) does not alter the value of the function P:

$$P(\mu q^A, p^A, \mu q^B, p^B) = P(q^A, p^A, q^B, p^B) \qquad (\mu \in \mathbb{R}_{++}).$$

(3.14) Mean Value Test:

The value of the function P always lies between the smallest and the largest price ratio of the corresponding prices:

$$\min\left\{\frac{p_1^B}{p_1^A}, \ldots, \frac{p_n^B}{p_n^A}\right\} \le P(q^A, p^A, q^B, p^B) \le \max\left\{\frac{p_1^B}{p_1^A}, \ldots, \frac{p_n^B}{p_n^A}\right\}.$$

Proof of Theorem (3.10):

The Identity Test (3.11) and the Weak Proportionality Test (3.12) immediately follow from the Proportionality Axiom (2.2). To prove the assertion in the case of (3.13), Axiom (2.3) implies

$$P(\mu q^A, p^A, \mu q^B, p^B) = P(\mu q^A, \lambda p^A, \mu q^B, \lambda p^B).$$

After setting $\mu = \frac{1}{\lambda}$ and applying the Commensurability Axiom (2.4) for $\lambda_1 = \lambda_2 = \ldots = \lambda_n = \lambda$

$$P\left(\frac{q^A}{\lambda}, \lambda p^A, \frac{q^B}{\lambda}, \lambda p^B\right) = P(q^A, p^A, q^B, p^B)$$

is derived.

All that remains is to prove the proposition for the Mean Value Test. We first define

$$a := \min\left\{\frac{p_1^B}{p_1^A}, \ldots, \frac{p_n^B}{p_n^A}\right\} \text{ and } b := \max\left\{\frac{p_1^B}{p_1^A}, \ldots, \frac{p_n^B}{p_n^A}\right\}.$$

Now, on one hand

$$a = P(q^A, p^A, q^B, ap^A) \leq P(q^A, p^A, q^B, p^B)$$

$$\text{(by (2.2))} \qquad\qquad \text{(by (2.1))}$$

and, on the other hand

$$b = P(q^A, p^A, q^B, bp^A) \geq P(q^A, p^A, q^B, p^B).$$

$$\text{(by (2.2))} \qquad\qquad \text{(by (2.1))}$$

Hence,

$$\min\left\{\frac{p_1^B}{p_1^A}, \ldots, \frac{p_n^B}{p_n^A}\right\} \leq P(q^A, p^A, q^B, p^B) \leq \max\left\{\frac{p_1^B}{p_1^A}, \ldots, \frac{p_n^B}{p_n^A}\right\}. \quad \square$$

Methods for constructing new price indexes or purchasing power parities from given ones are described in the following three theorems. The proof of Theorem (3.15) is obvious and the proofs for Theorems (3.18) and (3.20) can be found in Krtscha [1979].

(3.15) Theorem:

If $P_1,...,P_k$ are price indexes/purchasing power parities in accordance with Definition (2.0), then

$$(\alpha_1 P_1^\delta +...+ \alpha_k P_k^\delta)^{1/\delta} \qquad \begin{cases} \delta \neq 0,\ \alpha_1 \geq 0,...,\alpha_n \geq 0 \\ \text{real constants, } \Sigma\alpha_i = 1 \end{cases} \qquad (3.16)$$

and

$$P_1^{\delta_1} \cdot P_2^{\delta_2} \cdot ... \cdot P_k^{\delta_k} \qquad \begin{cases} \delta_1 \geq 0,...,\delta_k \geq 0, \\ \text{real constants, } \Sigma\delta_i = 1 \end{cases} \qquad (3.17)$$

also represent price indexes/purchasing power parities. P^δ is defined as

$$(q^A,p^A,q^B,p^B) \vdash [P(q^A,p^A,q^B,p^B)]^\delta.$$

(3.18) Theorem:

If $P_1,...,P_k$ are price indexes/purchasing power parities in accordance with Definition (2.0), then the function defined by

$$(q^A,p^A,q^B,p^B) \vdash \bar{P}(P_1(q^A,p^A,q^B,p^B),...,P_k(q^A,p^A,q^B,p^B)) \qquad (3.19)$$

is also a price index/purchasing power parity if $\bar{P}: \mathrm{IR}^k_{++} \to \mathrm{IR}_{++}$ is linearly homogenous, monotonically increasing and $\bar{P}(1,...1) = 1$.

(3.20) Theorem:

If P is a price index/purchasing power parity according to Definition (2.0), then

$$P(a_1 q_1^A,...,a_n q_n^A, b_1 p_1^A,...,b_n p_n^A, c_1 q_1^B,...,c_n q_n^B, b_1 p_1^B,...,b_n p_n^B) \qquad (3.21)$$

$$(a_1,...,a_n, b_1,...,b_n, c_1,...,c_n \text{ positive real constants})$$

also represents a price index/purchasing power parity.

We point out here that a procedure or a set of procedures that generates **all** functions P satisfying Axioms (2.1) to (2.4), i.e., the set of all mechanistic (statistical) price indexes, is not yet known.

It was the objective of this section to throw some light on the reasonableness of Definition (2.0). It must be admitted, however, that the class of functions obeying (2.0) is still extremely wide, for an infinite number of functions $P: IR_{++}^{4n} \rightarrow IR_{++}$ can also be generated according to Theorems (3.15), (3.18), (3.20) and other procedures. Therefore, in the next section we go one step further and present additional conditions (tests, criteria) which, required in addition to Axioms (2.1) to (2.4), may drastically reduce the number of functions conforming to the new set of conditions. Later, in Section 5, we will look at those sets of axioms/tests where only one particular price index/purchasing power parity or just one class of functions P satisfies the respective set of requirements.

4. Further Criteria and Implications

Most of the requirements now mentioned are again stated as functional equations which are valid for all (q^A, p^A, q^B, p^B):

(4.1) Linear Homogeneity Test:

If all comparison time/country prices change λ-fold ($\lambda \in IR_{++}$), then the value of the function P is changed by λ:

$$P(q^A, p^A, q^B, \lambda p^B) = \lambda P(q^A, p^A, q^B, p^B) \qquad (\lambda \in \mathrm{IR}_{++}).$$

(4.2) Theorem:

Every function $P: \mathrm{IR}_{++}^{4n} \to \mathrm{IR}_{++}$ satisfying the Linear Homogeneity Test (4.1) and the Identity Test (3.11) also fulfills the Proportionality Axiom (2.2). The converse is not true.

Proof:

After setting $p^B = p^A$, Tests (4.1) and (3.11) imply the Proportionality Axiom (2.2). That the converse does not hold is verified by the following function:

$$P(q^A, p^A, q^B, p^B) = \frac{q^A(p^A + p^B)}{q^A p^A} \cdot \frac{q^B p^B}{q^B(p^A + p^B)} \qquad (4.3)$$

which meets the Proportionality Axiom (2.2) and, hence, the Identity Test (3.11), but not the Linear Homogeneity Test (4.1). $\square$

In addition, the function given by (4.3) also satisfies the Monotonicity Axiom (2.1), the Price Dimensionality Axiom (2.3), and the Commensurability Axiom (2.4). Hence, Corollary (4.4) is immediately deduced:

(4.4) Corollary:

The class of functions $P: \mathrm{IR}_{++}^{4n} \to \mathrm{IR}_{++}$ conforming to Axioms (2.1) to (2.4) is wider than the class of functions satisfiying Axioms (2.1), (2.3), (2.4), the Identity Test (3.11) and the Linear Homogeneity Test (4.1).

In our monograph (Eichhorn/Voeller [1976]; see also Eichhorn [1976, 1978]) we used Axioms (2.1), (2.3), (2.4) and Tests (3.11), (4.1) to define price indexes.

Next, four tests are considered three of which impose a kind of reversal procedure on a function P: $\mathrm{IR}^{4n}_{++} \to \mathrm{IR}_{++}$. Three of the tests, that is, the Time/Country Reversal Test, the Factor Reversal Test and the Product Test (Weak Factor Reversal Test) originally date back to I. Fisher's investigations [1927] while the Price Reversal Test was first published in Funke/Voeller [1979]. All four tests deserve attention since they will be used to characterize a certain price index/purchasing power parity.

(4.5) Time/Country Reversal Test:

If P (q^A, p^A, q^B, p^B) compares quantities and prices of comparison time/country B with those of base time/country A and $P(q^B, p^B, q^A, p^A)$ does the same in the opposite direction, then the product of the values of the two function values equals one:

$$P(q^A, p^A, q^B, p^B) \cdot P(q^B, p^B, q^A, p^A) = 1.$$

The fulfillment of this test implies that it does not matter which time/country is taken as denominator time/country ("base time/country invariance"). One index P can be easily calculated as the reciprocal of the other. Kravis *et al.* [1975, p.47] comment on this requirement, referring only to interspatial comparisons, as follows: "In many applications the utility of the results would be greatly diminished if, for example, it were necessary to provide two estimates of the comparison between the countries depending upon which was taken as the denominator country in the ratio between them."

(4.6) Factor Reversal Test:

In P, q^A and p^A as well as q^B and p^B are interchanged. The resulting $P(p^A, q^A, p^B, q^B)$ can be regarded as the value of a quantity index if $P(q^A, p^A, q^B, p^B)$ is the value of a price index/purchasing power parity.[4] The product of the two values equals the expenditure ratio (value ratio) for the two baskets of goods:

$$P(q^A, p^A, q^B, p^B) \cdot P(p^A, q^A, p^B, q^B) = \frac{q^B p^B}{q^A p^A}.$$

A weaker version of Test (4.6) represents the Product Test:

(4.7) **Product Test**:

The product of the values of P and of a quantity index Q: $IR_{++}^{4n} \rightarrow IR_{++}$ which satisfies axioms analogous to (2.1) to (2.4) equals the expenditure ratio:

$$P(q^A, p^A, q^B, p^B) \cdot Q(q^A, p^A, q^B, p^B) = \frac{q^B p^B}{q^A p^A}.$$

On the practical use of the Product Test Drechsler [1973, p.22] may be quoted: "One should not forget here that in many cases quantity indexes cannot be compiled directly but only by deflation of value data by means of price indexes." Contrary to this procedure the price index/purchasing power parity and the quantity index must be computed independently in order that the stronger Factor Reversal Test (4.6) is satisfied.

A rather controversial test is given by

(4.8) **Price Reversal Test**:

If all base and comparison time/country prices are interchanged but the quantities remain unchanged, then the product of the values of the two function values equals one:

$$P(q^A, p^A, q^B, p^B) \cdot P(q^A, p^B, q^B, p^A) = 1.$$

For critical comments on this condition which, by the way, is satisfied by such famous indexes like the Laspeyres index (3.1), the Paasche index (3.2), and Fisher's "ideal index" (3.3) see Sato [1980] and Funke/Voeller [1980].

The next requirement contains a rather weak continuity assumption:

(4.9) Determinateness Test:

If any scalar argument in P tends to zero, then the value of the function P tends to a unique positive real number.

Now, two new criteria are introduced.

$$P(q^A, p^A, q^B, \lambda_1 p_1^A, \ldots, \lambda_n p_n^A) \tag{4.10}$$

$$= P(q^A, \lambda_1 p_1^A, \ldots, \lambda_n p_n^A, q^B, \lambda_1^2 p_1^A, \ldots, \lambda_n^2 p_n^A)$$

$$(\lambda_i \in \mathbb{R}_{++}, i = 1, \ldots, n).$$

In words, Condition (4.10) says:

As long as the price of each commodity in the base and comparison time/country differs by the same factor λ_i ($\lambda_i \in \mathbb{R}_{++}$, $i = 1, \ldots, n$), the value of the function P remains the same.

A somewhat stronger version of Condition (4.10) is given by

$$P(q^A, p^A, q^B, \lambda_1 p_1^A, \ldots, \lambda_n p_n^A) = P(q^A, p^C, q^B, \lambda_1 p_1^C, \ldots, \lambda_n p_n^C) \tag{4.11}$$

$$(\lambda_i \in \mathbb{R}_{++}, i = 1, \ldots, n).$$

Incidentally, Conditions (4.10) and (4.11) imply that the weight that each commodity is given in the basket of goods does not change in the course of further comparisons. The often used indexes denoted by (3.1), (3.2), (3.3) and many other price indexes/purchasing power parities do not conform to this requirement while the Cobb-Douglas Index (3.9) does. A detailed description of the implications of Property (4.11) taken together with other conditions is postponed until Section 5. At this point just one notable consequence of Condition (4.11) is evaluated:

(4.12) Remark:

Conformance to Condition (4.11) by a function P implies that with respect to the prices the function P only depends on the price relatives for the n commodities.

Proof:

Applying (4.11),

$$P(q^A, p^A, q^B, p^B) = P(q^A, p^A, q^B, \frac{p_1^B}{p_n^A} p_1^A, \ldots, \frac{p_n^B}{p_n^A} p_n^A)$$

$$= P(q^A, 1, \ldots, 1, q^B, \frac{p_1^B}{p_1^A}, \ldots, \frac{p_n^B}{p_n^A}). \quad \square$$

The presentation of tests or conditions used for intertemporal or interregional comparison methods would be incomplete if the famous Circular Test as well as its implications were not considered thoroughly. In his pathbreaking book "The Making of Index Numbers" I. Fisher commented on the "merit" of the Circular Test by saying [1927, p.271]: "I aim to show that the Circular Test is theoretically a mistaken one, that a irreducible minimum of divergence from such fulfillment is entirely right and proper, and, therefore, that a perfect fulfillment of this so-called Circular Test should really be taken as proof that the formula which fulfills it is erroneous." It is not surprising that Fisher's rigorous statement has stirred up many controversies among statisticians. By and large they objected to Fisher's conclusion and generally considered the failure of any purchasing power parity (or, equivalently, any price index formula) to pass the Circular Test as a shortcoming of this particular index. As a result, the Circular Test is given credit in almost all index-theoretic publications. Especially the further development and increased application of multilateral purchasing power parities have generated a number of statistical methods satisfying the Circular Test. Thus, a clash of opinion with Fisher's assertion seems to be inevitable. A concluding answer to his contention will not be given until Section 5.

(4.13) **Circular Test**:

If (q^A, p^A, q^B, p^B) represents the value of a price index/purchasing power parity between times/locations A and B and $P(q^B, p^B, q^C, p^C)$ represents the value for a comparison between times/locations B and C, then the product of the two values of P equals the value of a price index/purchasing power parity between times/places A and C:

$$P(q^A, p^A, q^B, p^B) \cdot P(q^B, p^B, q^C, p^C) = P(q^A, p^A, q^C, p^C).$$

In many publications the Circular Test (4.13) is also called the "chain method" since the separate time-to-time or place-to-place comparisons are joined together by successive multiplications like the links of a chain. Sometimes, the term "transitivity" requirement is used, too.

A weakened version of the Circular Test is given by the Base Test whose economic meaning is essentially the same as that of (4.13). Formally, the requirements are weaker, however:

(4.14) **Base Test**:

There exist functions P and $\tilde{P}$ such that either

$$P(q^A, p^A, q^B, p^B) \cdot \tilde{P}(q^B, p^B, q^C, p^C) = P(q^A, p^A, q^C, p^C)$$

or

$$\tilde{P}(q^A, p^A, q^B, p^B) \cdot P(q^B, p^B, q^C, p^C) = P(q^A, p^A, q^C, p^C)$$

holds.

Note that $\widetilde{P}$ only depends on four of the six vectors occurring in (4.14).

Now it is trivial to prove

(4.15) Remark:

Every function P: $IR_{++}^{4n} \to IR_{++}$ satisfying the Identity Test (3.11) and the Circular Test (4.13) also satisfies the Time/Country Reversal Test (4.5).

The Circular Test (4.13) and the Time/Country Reversal Test (4.5) lead to the following version of the Circular Test where over a number of intermediate steps a closed circuit is achieved:

$$P(q^A,p^A,q^B,p^B) \cdot P(q^B,p^B,q^C,p^C) \cdot \ldots \cdot P(q^K,p^K,q^A,p^A) = 1. \qquad (4.16)$$

From Funke/Hacker/Voeller [1979, p.682] the following theorem is taken:

(4.17) Theorem:

Every function P: $IR_{++}^{4n} \to IR_{++}$ satisfying the Linear Homogeneity Test (4.1) and the Circular Test (4.13) also conforms to the Price Dimensionality Axiom (2.3). The converse is not true.

Without proof as well the next theorem is cited from Eichhorn/Voeller [1976, p.33]:

(4.18) Theorem:

Every function P: $IR_{++}^{4n} \to IR_{++}$ fulfilling the weak Proportionality Test (3.12) and the Base Test (4.14) also satisfies the Circular Test (4.13). The converse is not true.

Finally, two variations of the Circular Test (4.13) are presented. Since the economic meaning of each condition can easily be grasped we just write down the respective functional equations:

(4.19) Price Circular Test:

$$P(q^X,p^A,q^Y,p^B) \cdot P(q^X,p^B,q^Y,p^C) = P(q^X,p^A,q^Y,p^C).$$

A special case of both the Circular Test (4.13) and the Price Circular Test (4.19) represents

(4.20) **Weak Circular Test**:

$$P(q^A,p^A,q^A,p^B) \cdot P(q^A,p^B,q^A,p^C) = P(q^A,p^A,q^A,p^C).$$

Now it is easy to confirm

(4.21) **Remark**:

Every function $P: \mathbb{R}^{4n}_{++} \to \mathbb{R}_{++}$ conforming to the Price Circular Test (4.19) also conforms to the Identity Test (3.11).

The following theorem, independently derived by Hacker [1979, p.77] and Krtscha [1979, p.70], provides important insight into the influence of the Circular Test (4.13) on the shape of an index function:

(4.22) **Theorem**:

Every function $P: \mathbb{R}^{4n}_{++} \to \mathbb{R}_{++}$ meeting the Identity Test (3.11) and the Circular Test (4.13) only depends on the base and comparison time/country prices but not on the quantities of the commodities.

Proof:

From (4.13) we obtain

$$P(q^A,p^A,q^B,p^B) = \frac{P(q^A,p^A,q^C,p^C)}{P(q^B,p^B,q^C,p^C)} .$$

Setting $q^C = (1,...,1)$ and $p^C = (1,...,1)$ it follows that

$$P(q^A,p^A,q^B,p^B) = \frac{f(q^A,p^A)}{f(q^B,p^B)}, \quad \text{where } f(q^A,p^A) = P(q^A,p^A,1,...,1).$$

Applying the Identity Test (3.11) yields

$$P(q^A,p^A,q^B,p^A) = \frac{f(q^A,p^A)}{f(q^B,p^A)} = 1,$$

hence,

$$f(q^A,p^A) = f(q^B,p^A).$$

Obviously f does not depend on q^A or q^B. Therefore we have

$$f(q^A,p^A) = g(p^A)$$

which leads to

$$P(q^A,p^A,q^B,p^B) = \frac{g(p^A)}{g(p^B)}. \quad \square$$

(4.23) **Corollary**:

If the Circular Test (4.13) is replaced by the Base Test (4.14) in Theorem (4.22), then the same implication holds.

The **proof** runs analogously to the proof of Theorem (4.22).

Since the Identity Test (3.11) is a special case of the Proportionality Axiom (2.2) the following corollary can be stated:

(4.24) **Corollary**:

Every function P: $IR_{++}^{4n} \to IR_{++}$ satisfying the Proportionality Axiom (2.2) and the Circular Test (4.13) (or the Base Test (4.14)) only depends on the prices.

Now it is extremely important to realize that Corollary (4.24) has severe consequences for the evaluation of several so-called multilateral purchasing power parities and also for the so-called chain methods in price index theory. **Since the fulfillment of the Circular Test (4.13) is the basic feature of all those methods and since almost all use quantity weights in their calculations they cannot conform to the Proportionality Axiom (2.2).** We point out here that this non-fulfillment represents a decisive drawback. Note that such an index cannot be regarded as a price index/purchasing power parity according to Definition (2.0).

5. Characterizations

The purpose of this section is to present characterizations of well-known price indexes/purchasing power parities. The search for characterizations is attributable not only to the intellectual challenge such a deduction represents but also to its great practical importance. For the statistician who is interested in knowing the class of functions satisfying a certain subset of requirements a characterization offers a definite answer. Characterizations of **cost-of-living indexes** are given by Diewert and Pollak in this volume.

Of course any given set of conditions should be independent in the following sense: no condition is deductible from the remaining ones. Hence, the given set of conditions is also **strict** (or minimal) which means that after dropping any one of the conditions the characterization and therefore the implications do not hold any more. Thus, redundant sets of conditions, i.e., sets containing conditions which are irrelevant to a characterization are not considered.

(5.1) **Theorem**:

The Monotonicity Axiom (2.1), the Proportionality Axiom (2.2), the Commensurability Axiom (2.4), and the Circular Test (4.13) (or the Base Test (4.14)) are independent in the following sense: Any three conditions can be satisfied by a function P: $\mathrm{IR}_{++}^{4n} \to \mathrm{IR}_{++}$ which does not satisfy the remaining condition.

Proof:

The function represented by (2.7) satisfies (2.2), (2.4) and (4.13) (hence (4.14)), but not (2.1). The function given by

$$P(q^A, p^A, q^B, p^B) = \sqrt{\frac{q^B p^B}{q^A p^A}} \qquad (5.2)$$

fulfills (2.1), (2.4) and (4.13) (hence (4.14)), but not (2.2).

The function given by

$$P(q^A, p^A, q^B, p^B) = \frac{cp^B}{cp^A} \qquad \text{(``Lowe index'',} \qquad (5.3)$$
$$c_1 > 0, \dots, c_n > 0 \text{ real constants)}$$

passes (2.1), (2.2) and (4.13) (hence (4.14)), but not (2.4). Finally, the function denoted by (3.1) conforms to (2.1), (2.2) and (2.4), but not to (4.13) (and not to (4.14)). $\square$

(5.4) Theorem:

A function P: $\mathrm{IR}_{++}^{4n} \to \mathrm{IR}_{++}$ fulfills the Monotonicity Axiom (2.1), the Proportionality Axiom (2.2), the Commensurability Axiom (2.4) and the Base Test (4.14) if and only if P is given by the Cobb-Douglas Index (3.9).

Proof (see Funke/Hacker/Voeller [1979]):

"$<=$": This direction is easily verified by inserting (3.9) into (2.1), (2.2), (2.4) and (4.14).

"$=>$": As proven in Eichhorn/Voeller [1976, p.35] a function P satisfying the Commensurability Axiom (2.4) and the Base Test (4.14) can be written in the following form:

$$P(q^A,p^A,q^B,p^B) = \frac{G(q_1^B p_1^B,...,q_n^B p_n^B)}{H(q_1^A p_1^A,...,q_n^A p_n^A)} \Phi(\frac{p_1^B}{p_1^A},...,\frac{p_n^B}{p_n^A}) \tag{5.5}$$

with functions $G,H,\Phi:\mathrm{IR}_{++}^n \rightarrow \mathrm{IR}_{++}$ and function Φ multiplicative, i.e.,

$$\Phi(\lambda_1\mu_1,...,\lambda_n\mu_n) = \Phi(\lambda_1,...,\lambda_n)\Phi(\mu_1,...,\mu_n). \tag{5.6}$$

Applying the Identity Test (3.11) as a special case of the Proportionality Axiom (2.2) to (5.5) one at once derives

$$\frac{G(q_1^B p_1^A,...,q_n^B p_n^A)}{H(q_1^A p_1^A,...,q_n^A p_n^A)} \Phi(1,...,1) = 1. \tag{5.7}$$

Setting first $p^A = (1,...,1)$ and $q^A = (1,...,1)$ and then $p^A = (1,...,1)$ and $q^B = (1,...,1)$, Equation (5.7) changes to

$$G(q_1^B,...,q_n^B) = c \; \epsilon \; \mathrm{IR}_{++}$$

and

$$H(q_1^A,...,q_n^A) = c \; \epsilon \; \mathrm{IR}_{++},$$

respectively. Hence

$$P(q^A,p^A,q^B,p^B) = \Phi(\frac{p_1^B}{p_1^A},...,\frac{p_n^B}{p_n^A}) .$$

Because of the Monotonicity Axiom (2.1) Φ is strictly increasing. This result and (5.6) imply (see Eichhorn [1978, p.66]) that Φ can be written as

$$\Phi\left(\frac{p_1^B}{p_1^A},...,\frac{p_n^B}{p_n^A}\right) = \left(\frac{p_1^B}{p_1^A}\right)^{\alpha_1} \cdots \left(\frac{p_n^B}{p_n^A}\right)^{\alpha_n}$$

where $\alpha_1,...,\alpha_n$ are positive real constants. The application of the Proportionality Axiom (2.2) finally yields $\Sigma\alpha_i = 1$, i.e., Formula (3.9). $\square$

(5.8) Remark:

If in Theorem (5.4) the Base Test (4.14) is replaced by the Circular Test (4.13), the same result as in Theorem (5.4) holds.

The **proof** runs analogously to that of Theorem (5.4).

With Remark (5.8) the ground is laid for a conclusive answer to the old controversy about the value of the Circular Test. On page 431, Irving Fisher's standpoint was briefly indicated and it is summed up again in the following sentence [1927, p.276]: "It is clear that constant weighting, though it makes it possible to fulfill the circular test, does so at the expense of forcing the facts, for the true weights are **not** constants."

What Fisher assumed without final proof, that is, that formulas which satisfy the Circular Test are certain index numbers with constant weights is now confirmed by the results of Theorem (5.4) and Remark (5.8). In both statements the α_is of function (3.9) can be interpreted as constant weights. Besides the Cobb-Douglas Index, Fisher also noticed Lowe's Index given by (5.3) as conforming to his Circular Test. But according to Definition (2.0) Lowe's Index is not a price index/purchasing power parity since it does not fulfill the Commensurability Axiom (2.4). Thus, the fact cannot be denied that the main purpose of the Circular Test, that is, the adjustment of the quantity weights to the new situation in each new dual comparison around a circle of points of time or countries cannot be accomplished. There simply does not exist a function P which satisfies our four basic axioms for price indexes/purchasing power parities and the Circular Test, simultaneously, **and** which enables adequately changing weights. In fact, only the Cobb-Douglas function (3.9) with its constant weighting scheme conforms to the basic requirements and the Circular Test which currently seems to be a *conditio sine qua non* specially in multilateral purchasing power parity comparisons.

Thus, the following remark can be taken as a late justification of I. Fisher's repudiation of the Circular Test:

(5.9) Remark:

The fulfillment of the Circular Test (4.13) or the Base Test (4.14) in all possible methods of intertemporal or interspatial price comparisons is achieved by:

(i) constant weighting schemes which contradict the "spirit" of the Circular or Base Test;

(ii) non-fulfillment of one or more of the four basic axioms given in Definition (2.0).

The conclusion reached in Remark (5.9) is the reason why Funke/Hacker/Voeller in [1979, p.686] proposed to discard the Circular Test and the Base Test and to use instead the two variations (4.19) and (4.20). In other words, in a circular price comparison it makes much more sense to do without the requirement of changing weights which cannot be met anyway.

A further characterization of the Cobb-Douglas Index is possible using the following requirements:

(5.10) Theorem:

The Monotonicity Axiom (2.1), the Proportionality Axiom (2.2), Condition (4.11), and the Circular Test (4.13) are independent in the following sense: Any three of these conditions can be satisfied by a function P: $IR_{++}^{4n} \rightarrow IR_{++}$ which does not conform to the remaining condition.

i.e., only depends on the prices. Because of Remark (4.12), Condition (4.11) implies

$$P(q^A, p^A, q^B, p^B) = P(q^A, 1, \ldots, 1, q^B, \frac{p_1^B}{p_1^A}, \ldots, \frac{p_n^B}{p_n^A}).$$

Hence,

$$\tilde{P}(\frac{p_1^B}{p_1^A}, \ldots, \frac{p_n^B}{p_n^A}) = \frac{\bar{P}(p_1^B, \ldots, p_n^B)}{\bar{P}(p_1^A, \ldots, p_n^A)}, \qquad (5.14)$$

where $\tilde{P}(\gamma) := \dfrac{f(\gamma)}{f(1, \ldots, 1)}$

After substituting

$$p_i^B = p_i^B p_i^A$$

we obtain

$$\tilde{P}(p_1^B, \ldots, p_n^B)\tilde{P}(p_1^A, \ldots, p_n^A) = \tilde{P}(p_1^B p_1^A, \ldots, p_n^B p_n^A), \qquad (5.15)$$

i.e., a multiplicative function of type (5.6). Because of the Monotonicity Axiom (2.1) $\tilde{P}$ is strictly increasing. This result and (5.15) imply (see Eichhorn [1978, p.66])

$$\tilde{P}(\frac{p_1^B}{p_1^A}, \ldots, \frac{p_n^B}{p_n^A}) = \left(\frac{p_1^B}{p_1^A}\right)^{\alpha_1} \left(\frac{p_2^B}{p_2^A}\right)^{\alpha_2} \cdots \left(\frac{p_n^B}{p_n^A}\right)^{\alpha_n}$$

with $\alpha_1 > 0, \ldots, \alpha_n > 0$ positive real constants. Now the application of the Proportionality Axiom (2.1) implies $\Sigma \alpha_i = 1$ and, thus, leads to the Cobb-Douglas Index (3.9). The other direction of the proof is obvious. $\square$

Proof:

The function given by (2.7) meets (2.2), (4.11) and (4.13), but not (2.1). The function represented by

$$P(q^A, p^A, q^B, p^B) = \left(\frac{p_1^B}{p_1^A}\right)^{\alpha_1} \left(\frac{p_2^B}{p_2^A}\right)^{\alpha_2} \cdots \left(\frac{p_n^B}{p_n^A}\right)^{\alpha_n} \tag{5.11}$$

$$(\alpha_1 > 0, \ldots, \alpha_n > 0 \text{ positive real constants}, \Sigma \alpha_i \neq 1)$$

satisfies (2.1), (4.11) and (4.13), but not (2.2). The function denoted by (2.10) fulfills (2.1), (2.2) and (4.13), but not (4.11). At last, the function given by

$$P(q^A, p^A, q^B, p^B) = \frac{1}{n} \Sigma \frac{p_i^B}{p_i^A} \tag{5.12}$$

conforms to (2.1), (2.2) and (4.11), but not to (4.13). □

(5.13) Theorem:

A function P: $IR_{++}^{4n} \rightarrow IR_{++}$ satisfies the Monotonicity Axiom (2.1), the Proportionality Axiom (2.2), Condition (4.11) and the Circular Test (4.13) if and only if P is the Cobb-Douglas Index (3.9).

Proof:

According to Remark (4.24) every function P conforming to the Proportionality Axiom (2.2) and the Circular Test (4.13) can be written as

$$P(q^A, p^A, q^B, p^B) = \frac{f(p^B)}{f(p^A)},$$

Many alternative aggregation methods have been favored by one statistical agency or another but only very few have been as widely used as Irving Fisher's so-called "ideal index" denoted by (3.3). It produces results that in every instance are intermediate between its two component indexes (3.1) and (3.2). Hence, it has been regarded as an evenhanded compromise that is often applied for pragmatic reasons: it serves the interests of brevity and it avoids the cumbersome presentation of all comparisons with the two more basic indexes.

Historically, Fisher [1927, p.220 ff.] called the function given by (3.3) "ideal" since it satisfied all of his tests for price indexes except for the Circular Test (4.13). As mentioned before, Fisher therefore proposed to discard the Circular Test. However, his assumed superiority of the "ideal index" over all other index numbers has often been challenged in the literature and, today, a more application-oriented approach seems to be predominant. A particular price index/purchasing power parity is chosen with the properties desirable for a special assignment. Characterizations which offer clear-cut answers to such problems therefore deserve increased attention.

The following characterization of the Fisher Index has been originally published in Funke/Voeller [1979]. To achieve it the three Reversal Tests (4.5), (4.6) and (4.8) are sufficient.

(5.16) **Theorem**:

The Time/Country Reversal Test (4.5), the Factor Reversal Test (4.6) and the Price Reversal Test (4.8) are independent in the following sense: any two of these conditions can be satisfied by a function P: $IR_{++}^{4n} \to IR_{++}$ which does not satisfy the remaining condition.

Proof:

The function given by

$$P(q^A, p^A, q^B, p^B) = \tag{5.16}$$

$$\left[\frac{q^A p^B \; q^B p^B}{q^A p^A \; q^B p^A}\right]^{\frac{1}{2}} \frac{(q^B + p^B)p^A}{(q^A + q^B)p^A} \; \frac{(q^B + p^A)q^A}{(q^B + p^B)q^A} \; \frac{(q^A + p^A)p^B}{(q^B + p^A)p^B} \; \frac{(q^A + p^B)q^B}{(q^A + p^A)q^B}$$

meets (4.6) and (4.8), but not (4.5). The function denoted by (3.7) conforms to (4.5) and (4.8), but not to (4.6). Finally, the function represented by (5.2) fulfills (4.5) and (4.6), but not (4.8). □

(5.17) **Theorem:**

A function P: $\mathbb{R}^{4n}_{++} \to \mathbb{R}_{++}$ satisfies the Time/Country Reversal Test (4.5), the Factor Reversal Test (4.6) and the Price Reversal Test (4.8) if and only if P is Fisher's "ideal index" (3.3).

Proof:

" $<=$ ": is obvious.

" $=>$ ": From (4.5)

$$P(q^B, p^B, q^A, p^A) = \frac{1}{P(q^A, p^A, q^B, p^B)}$$

and from (4.8)

$$P(q^A, p^B, q^B, p^A) = \frac{1}{P(q^A, p^A, q^B, p^B)}$$

the following equation follows:

$$P(q^A, p^B, q^B, p^A) = P(q^B, p^B, q^A, p^A) \quad 5 \tag{5.18}$$

Interchanging p^A and p^B in (4.6) yields

$$P(q^A,p^B,q^B,p^A) \cdot P(p^B,q^A,p^A,q^B) = \frac{q^B p^A}{q^A p^B} . \tag{5.19}$$

Using (5.18), Equation (5.19) changes into

$$P(q^B,p^B,q^A,p^A) \cdot P(p^A,q^A,p^B,q^B) = \frac{q^B p^A}{q^A p^B} . \tag{5.20}$$

Dividing (4.6) by (5.20) results in

$$\frac{P(q^A,p^A,q^B,p^B)}{P(q^B,p^B,q^A,p^A)} = \frac{q^B p^B}{q^A p^A} \frac{q^A p^B}{q^B p^A}. \tag{5.21}$$

The multiplication of (5.21) with (4.5) finally yields

$$[P(q^A,p^A,q^B,p^B)]^2 = \frac{q^A p^B}{q^B p^A} \frac{q^B p^B}{q^A p^A},$$

i.e., Fisher's "ideal index" (3.3). □

6. Inconsistency Considerations

In Eichhorn/Voeller [1976] we presented the general solution to the so-called inconsistency problem of Fisher's tests. Since these tests form a subset of all the conditions considered in this investigation all previous results concerning inconsistent subsets of conditions are also valid here. But since new conditions have been added new inconsistencies may turn up, too. In the following a very definite viewpoint is taken with respect to these potential inconsistencies. We only look for those inconsistent subsets which result when additional conditions are added to a price index/purchasing power parity of type (2.0). This means, that we assume the fulfillment of Axioms (2.1) to (2.4) (and, hence, according to Remark

(3.10) of Tests (3.11), (3.12), (3.13), (3.14)) by a function P: $IR_{++}^{4n} + IR_{++}$ and then ask which further conditions produce an inconsistent set of properties. Therefore we speak of an **inconsistency theorem** or a **non-existence theorem** in connection with an index function of type (2.0) and all other conditions (tests) listed additionally in Sections 3 and 4 if for a subset of these conditions there does not exist a function P of type (2.0) which satisfies all conditions of the subset at the same time.

(6.1) Remark:

Inconsistency theorems are at best possible if a condition from $\left\{ \right.$(4.10), (4.11), (4.13), (4.14)$\left. \right\}$ and a condition from $\left\{ \right.$(4.6), (4.7), (4.9)$\left. \right\}$ are added to Axioms (2.1) to (2.4).

Proof:

Fisher's "ideal index" (3.3) satisfies all conditions except for (4.10), (4.11), (4.13) and (4.14). The Cobb-Douglas Index (3.9) meets all conditions except for (4.6), (4.7) and (4.9). □

Now, the number of possibly inconsistent sets of conditions for an index of type (2.0) is greatly reduced. As a further consequence of the characterizations of Fisher's "ideal index" (3.3) as well as the Cobb-Douglas Index (3.9) it is clear that inconsistencies occur when conditions are added to (2.1) to (2.4) which either index does not fulfill. In particular this implies:

(6.2) Theorem:

There does not exist any price index/purchasing power parity of type (2.0) which satisfies either:

(i)   the Base Test (4.14) (or the Circular Test (4.13)) and the Determinateness Test (4.9)

or

(ii) the Base Test (4.14) (or the Circular Test (4.13)) and the Product Test (4.7) (or the Factor Reversal Test (4.6))

simultaneously.

Proof:

According to Theorem (5.4), the only function $P: IR_{++}^{4n} \rightarrow IR_{++}$ fulfilling (2.1) to (2.4) and the Base Test (4.14) (or the Circular Test (4.13)) is given by (3.9). This index, however, does not satisfy simultaneously any of the sets of conditions listed under (i) and (ii). □

Of course, any additional condition added to (i) or (ii) also leads to inconsistencies.

(6.3) Theorem:

There does not exist any price index/purchasing power parity of type (2.0) which satisfies either:

(i) the Time/Country Reversal Test (4.5), the Factor Reversal Test (4.6), the Price Reversal Test (4.8) and Condition (4.10) (or Condition (4.11))

or

(ii) the Time/Country Reversal Test (4.5), the Factor Reversal Test (4.6), the Price Reversal Test (4.8) and the Base Test (4.14) (or the Circular Test (4.13))

simultaneously.

Proof:

According to Theorem (5.17) Fisher's "ideal index" (3.3) can be characterized by Conditions (4.5), (4.6) and (4.8). Since (3.3) does not fulfill the above combinations the theorem holds. □

Footnotes

* We would like to thank Bert Balk and Erwin Diewert for useful comments.

1 Please note the following notation: $x = (x_1,...,x_n) > (y_1,...,y_n) = y$ if $x_1 > y_1,...,x_n > y_1$, and $x \geq y$ if $x_1 \geq y_1,...,x_n \geq y_n$ but $x \neq y$, and $x \geqq y$ if $x_1 \geq y_1,...,x_n \geq y_n$.

2 xy always denotes the inner product of the two vectors x and y.

3 From now on the term "index" is used equivalently to "price index" and/or "purchasing power parity".

4 A quantity index is required to satisfy conditions analogous to those satisfied by a price index/purchasing power parity.

5 As indicated in Footnote 4 the analogous condition for a quantity index would be given by

$$P(p^A, q^B, p^B, q^A) = P(p^B, q^B, p^A, q^A)$$

or, interchanging q^A and q^B, by

$$P(p^A, q^A, p^B, q^B) = P(p^B, q^A, p^A, q^B).$$

References

Diwert, W.E. [1983]. "The Theory of the Cost-of-Living Index and the Measurement of Welfare Change", *Price Level Measurement: Proceedings From a Conference Sponsored by Statistics Canada.*

Drechsler, L. [1973]. "Weighting of Index Numbers in Multilateral International Comparisons", *The Review of Income and Wealth*, pp.17-34.

Eichhorn, W. [1976]. "Fisher's Tests Revisited", *Econometrica* 44, pp.247-256.

———— [1978]. "Functional Equations in Economics", Addison-Wesley Pub. Co., Reading, Mass.

———— and J. Voeller [1976]. "Theory of the Price Index. Fisher's Test Approach and Generalizations", Lecture Notes in Economics and Mathematical Systems. Vol. 140, Springer Verlag, Berlin-Heidelberg-New York.

Fisher, I. [1922]. "The Making of Index Numbers", Houghton-Mifflin, Boston, Third edition revised, 1927. Reprinted by A.M. Kelley, New York, 1968.

Funke, H. and J. Voeller [1979]. "Characterization of Fisher's 'Ideal Index' by Three Reversal Tests", Statistische Hefte, 20. Jg., Hefte 1, pp.54-60.

———— [1980]. "Three Comments on Sato's Paper: The Price Reversal Test and Economic Indices", Statistische Hefte, 21. Jg., Hefte 2, pp.131-132.

———— G. Hacker and J. Voeller [1979]. "Fisher's Circular Test Reconsidered", Schweizerische Zeitschrift für Volkswirtschaft und Statistik, Hefte 4, pp.677-688.

Hacker, G. [1979]. "Theorie der wirtschaftlichen Kennzahl", Dissertation Karlsruhe.

Kravis, I.B. et al. [1975]. "A System of International Comparisons of Gross Product and Purchasing Power", Johns Hopkins University Press, Baltimore and London.

_____ [1978]. "International Comparisons of Real Product and Purchasing Power", Johns Hopkins University Press, Baltimore and London.

Krtscha, M. [1979]. "Über Preisindices und Deren Axiomatische Characterisierung", Dissertation Karlsruhe.

Pollak, R.A. "The Theory of the Cost-of-Living Index", *Price Level Measurement: Proceeding From a Conference Sponsored by Statistics Canada.*

Sato, K. [1980]. "The Price Reversal Test and Economic Indices", Statistische Heft, 21. Jg., Hefte 2, pp.127-132.

Voeller, J. [1982]. "Purchasing Power Parities for International Comparisons", Oelgeschlager, Gunn & Hain, Meisenheim-Cambridge.

PRICE LEVEL MEASUREMENT – W.E. Diewert (Editor)
Canadian Government Publishing Centre /
Elsevier Science Publishers B.V. (North-Holland)
© Minister of Supply and Services Canada, 1990

COMMENTS

Charles Blackorby
Department of Economics
University of British Columbia

There are two main strands of thought in the study of index numbers. The first is the so-called axiomatic method of constructing index numbers. In this particular approach the researcher considers an index number to be a function of current and base period prices and quantities, and then imposes an axiomatic structure on these index numbers. The result is a narrower class of index numbers than the original set. The other strand of thought approaches index numbers in a similar way except that the quantities, both current and base, are thought to be functions of the prices. Hence, in this alternative, which I'll refer to as the economic theory of index numbers, ultimately the index numbers are functions only of current and base period prices.

In the paper written by Professors Eichhorn and Joachim Voeller we see the high standards which are regularly found in the work of Professors Eichhorn and Voeller, two of the foremost experts in the world of the axiomatic theory of index numbers. In support of the exact economic index number approach, we have papers by Robert Pollak and Erwin Diewert on the Foundations of Economic Index Number Theory. These latter two papers represent the best that the other side has to offer.

Let me be more explicit about the difference between these two approaches. For Professors Eichhorn and Voeller a price index is a function which depends upon four sets of independent variables, that is, current and base period prices and current and base period quantities. This price index is assumed to satisfy four axioms, monotonicity, proportionality, price dimensionality and commensurability. It is important to notice that each of these four arguments is allowed to range independently over the entire positive orthant of Euclidean N-space.

To convert this to an economic index number, à la Diewert and Pollak, assume that the current quantities are the compensated demand functions of a utility maximizing household evaluated at current prices and a specified utility level, and that base period quantities are the compensated demands of the same household at base period prices and the same utility level. Thus, the index numbers in the exact economic sense are functions only of current and base period prices and a specified utility level (and, if you like, the preference ordering of the household in question).

Let me now return to the paper of Professors Eichhorn and Voeller. This paper is a paradigm of excellence in clarity of presentation, mathematical elegance and scholarship. The results are, however, essentially impossibility results. That is, suppose we restrict our attentions to the class of index numbers mentioned above. Then Eichhorn and Voeller show that if in addition to the above usual regularity assumptions, we require that this price index function satisfy a circularity test and a product test, then the class of possible price indices is empty. That is, there does not exist an index number which satisfies monotonicity, proportionality, price dimensionality, commensurability, circularity and the product test. The results in their paper are correct. In addition Professors Eichhorn and Voeller are very persuasive in the reasonableness of the axioms which they are maintaining. It also provides guidelines for the practical construction of index numbers.

Nevertheless this type of exercise puts me in a rather "whodunit" mood. Even though it is the entire set of axioms which is mutually inconsistent, I feel there must be some particular culprit. My candidate is the product test. Not only are prices and quantities independent variables, but the price index and the quantity index are independent variables too. Yet the product test says that they can't be independent.

The two positions outlined above are polar extremes precisely because of the domains of definitions of the index numbers in question. In the case of the mechanistic or axiomatic approach to index numbers, current quantities are independent of all prices and base period quantities are independent of all prices; it is merely the index number itself that must satisfy certain axioms. In the economic index number case current quantities are constrained to be functions of prices which satisfy the Slutsky symmetry conditions by adding up conditions of a single utility maximizing consumer.

What puzzles me about the two approaches, is not that researchers have investigated in great detail these polar cases, but rather that no one has investigated the middle ground. For example, suppose that we begin with Professor Eichhorn's definition of an index number. If we were to relax the assumption that quantities are completely unrelated, independent of prices, we do not need immediately assume that these quantities are the compensated demand functions of a single utility maximizing consumer. It might be sufficient to assume that current quantities are simply the aggregate demand functions of the economy which we are examining. We might then begin to place regularity conditions on these demand functions. The first step would be to assume the demand functions were continuous, homogeneous of degree zero and add to total expenditure. From a general equilibrium system, we can expect no more. We could also assume that the system of aggregate demand equations satisfy the assumption of universal gross substitutes. This, in conjunction with the axioms which Professor Eichhorn finds plausible, would lead to a smaller class of reasonable cost-of-living indices. We have no idea how small such a class might be. However, we do know that the assumption of gross substitutability is sufficient to ensure the uniqueness of competitive equilibrium as well as its global stability. We could go even further without assuming that the quantities in question are generated by those of a utility maximizing consumer. For example, it is known that aggregate demand functions may satisfy some Slutsky symmetry-like conditions but not all of them (under some circumstances); see for example a paper by E. Diewert [1977]. In conjunction with the above-mentioned axioms and some Slutsky symmetry-like conditions, the class of feasible index numbers gets larger. Precisely where to stop in this endeavour is by no means obvious, but if we were to systematically study this procedure of making the demand functions satisfy more and more restrictions, thereby enlarging the class of reasonable index numbers, we might learn how to interact with the kinds of demand functions which are actually generated by the real world.

Reference

Diewert, W.E., [1977] " Generalized Slutsky Conditions of Aggregate Consumer Demand",
    Journal of Economic Theory, 15, pp.353-362.

NAME INDEX

SUBJECT INDEX